I0031171

Understanding Conversational AI

Philosophy, Ethics, and Social Impact of Large Language Models

Thierry Poibeau

]u[

ubiquity press
London

Published by
Ubiquity Press Ltd.
Unit 3N, 6 Osborn Street
London, E1 6TD
United Kingdom
www.ubiquitypress.com

Text © Thierry Poibeau 2025

First published 2025

Cover design by Tom Grady
Cover image: *Detail from Paris Street;
Rainy Day, 1877 by Gustave Caillebotte (1848–1894)*
Cover image credit: Charles H. and Mary F.S. Worcester Collection,
Art Institute of Chicago. CCO Public Domain.
Message bubbles by juicy_fish on Freepik.com

Print and digital versions typeset by Siliconchips Services Ltd.

ISBN (Paperback): 978-1-914481-70-3
ISBN (PDF): 978-1-914481-71-0
ISBN (EPUB): 978-1-914481-72-7
ISBN (Mobi): 978-1-914481-73-4

DOI: https://doi.org/10.5334/bde

This work is licensed under the Creative Commons Attribution-
NonCommercial 4.0 International (CC BY-NC 4.0) License (unless
stated otherwise within the content of the work). To view a copy of this
license, visit https://creativecommons.org/licenses/by-nc/4.0/ or send
a letter to Creative Commons, 444 Castro Street, Suite 900, Mountain
View, California, 94041, USA. This license allows for copying and
distributing the work, providing author attribution is clearly stated and
that you are not using the material for commercial purposes.

The full text of this book has been peer-reviewed to ensure high academic
standards. For full review policies, see http://www.ubiquitypress.com/

Suggested citation:
Poibeau, T. 2025. *Understanding Conversational AI: Philosophy, Ethics,
and Social Impact of Large Language Models*. London: Ubiquity Press.
DOI: https://doi.org/10.5334/bde. License: CC BY-NC 4.0

To read the free, open access version of this
book online, visit https://doi.org/10.5334
/bde or scan this QR code with your mobile
device:

In some ways, [artificial intelligence] is akin to medieval alchemy.
We are at the stage of pouring together different combinations of
substances and seeing what happens, not yet having developed
satisfactory theories.

Winograd, T. (1977). On some contested suppositions
of generative linguistics about the scientific study of languages.
Cognition, 5.

Contents

Acknowledgments

The author wishes to express sincere gratitude to Pierre Beckmann, Xiangling Zhang and a third anonymous reviewer for their insightful and constructive feedback. Their detailed and thoughtful comments greatly contributed to improving the quality and clarity of the manuscript. The author is also deeply thankful to his PhD students, postdoctoral researchers, colleagues, and project partners, whose discussions and collaborations have been invaluable. Many of the ideas developed in this book emerged from these rich and stimulating exchanges. He further wishes to acknowledge with gratitude the administrative support within his laboratory, which has greatly contributed to creating the conditions that made this work possible. Finally, the author wishes to thank the editorial team at Ubiquity Press for their professionalism and dedication in overseeing the publication of this volume.

Funding

This book was published thanks to the support of PRAIRIE-PSAI (Paris Artificial Intelligence Research Institute–Paris School of Artificial Intelligence, reference ANR-22-CMAS-0007), of a Sophie Germain fellowship for exchanges between Paris Sciences et Lettres and the University of Cambridge (2024–2025), and of the ANR Médialex project (2021–2025, reference ANR-21-CE38-0016).

Bio

Thierry Poibeau is a CNRS researcher based in Paris at the École normale supérieure. He is also a fellow of PRAIRIE-PSAI (Paris Artificial Intelligence Research Institute – Paris School of Artificial Intelligence). His research focuses primarily on natural language processing, computational humanities, and the social impact of artificial intelligence. ORCID: https://orcid.org/0000-0003-3669 -4051

Companion website

A companion website for the book is available at: https://tpoibeau .github.io/ucai/

Speaking machines, thinking humans

Human language, once regarded as a uniquely human capacity, is now being reshaped through the capabilities of machines. Large language models (LLMs) such as ChatGPT, Mistral, and Lama, do not simply process language—they generate it, at scale and in real time, with striking fluency and coherence. For the first time, we are engaging with systems that can produce extended and meaningful text across a range of communicative forms, challenging long-held assumptions about the boundaries of human expression.

Trained on massive corpora of text drawn from books, websites, conversations, and code, LLMs use statistical patterns to generate linguistic sequences in response to prompts. While their outputs often appear meaningful, coherent, and even insightful, it remains an open question whether, or in what sense, they 'understand' language. Their capacity to simulate communication raises deep questions, not only about the nature of these models, but about the language and intelligence itself.

This book takes seriously the philosophical implications of this transformation. It asks not just what LLMs do but also what they mean. What does it say about language, thought, and knowledge

How to cite this book chapter:
Poibeau, T. 2025. *Understanding Conversational AI: Philosophy, Ethics and Social Impact of Large Language Models*. Pp. 1–10. London: Ubiquity Press. DOI: https://doi.org/10.5334/bde.a. License: CC BY-NC 4.0

that machines can now perform these tasks so convincingly? What are we to make of the fact that LLMs learn without necessarily understanding, predict without believing, respond without intending? And what kind of ethical, political, and epistemic world are we building when we integrate such models into our communications, institutions, and infrastructures?

1 Why philosophy matters now

Philosophy is often slow to respond to technological change. But LLMs demand careful and timely reflection—not because they represent a singular rupture but because they unsettle the categories through which we interpret language, cognition, and meaning. They challenge what it means to know, to learn, to understand, and to speak. They reopen long-standing debates in the philosophy of language and mind, while also raising new questions about responsibility, authorship, and the construction of knowledge.

This book does not attempt to define once and for all what LLMs are: whether they think, understand, hallucinate, or merely simulate. Rather, it explores how they function as complex, distributed artifacts whose significance lies as much in how they are used as in how they are built. LLMs are not autonomous agents but sociotechnical systems shaped by human prompts, institutional aims, and cultural imaginaries. Their meaning is not only statistical but social.

The approach taken here is philosophical in orientation: conceptual, critical, and interdisciplinary. While grounded in traditions from the philosophy of language, mind, and ethics, the analysis also draws on insights from cognitive science, science and technology studies, and media theory. The aim is not to settle debates but to map the terrain of questions we now face— and to offer tools for thinking clearly and critically about technologies that are changing how we communicate and what we take language to be.

The book addresses a fundamental philosophical issue posed by LLMs: How do LLMs 'know' what they know, and what kind of knowledge is this? What is the relationship between statistical pattern recognition and meaning? Can a language model be said to understand or intend? What are the ethical and political implications of generating language at scale? How do LLMs shape our epistemic environment, and who gets to define the boundaries of truth and sense? What happens to creativity, authorship, and the self in a world where machines contribute to language production?

Rather than offering definitive answers, this book invites the reader into a space of critical inquiry. It is not a guide to using LLMs, or a warning against them, or a celebration of their potential. It is a philosophical investigation—open-ended, reflective, and attentive to the unfamiliar challenges of the present moment.

In an age when language generation is no longer the exclusive province of human authors, we must ask anew: What is language, what is intelligence, and what is at stake in their intersection? This book begins there.

2 Content and structure of the book

The chapters of this book offer a sustained philosophical investigation into the emergence, function, and societal implications of LLMs. While the technological capabilities of these systems have garnered widespread attention, our aim is to understand what their existence and operation reveal about the nature of language, cognition, and human interaction in an increasingly mediated world. Each chapter addresses a distinct but interrelated set of questions, drawing from traditions in linguistics, philosophy of mind, ethics, and critical theory to make sense of the phenomenon of LLMs not merely as tools but as conceptual artifacts that challenge long-held assumptions.

The first part of the book, titled 'Sense without Reference,' examines how LLMs redefine our understanding of language, exploring their structure, capabilities, and limitations in the apparent absence of grounding or referential meaning.

Chapter 1 explores how these systems, trained solely on textual data and operating without perception or intention, enact a model of language grounded in internal pattern recognition rather than external reference. This perspective resonates, perhaps surprisingly, with structuralist and poststructuralist theories of language that emphasize differential meaning, discursive conventions, and the autonomy of linguistic systems. The chapter sets the stage for viewing LLMs not only as engineering feats but also as "philosophical provocations" (McGinn, 2017). that demand a reassessment of how meaning is constituted in the absence of lived experience.

Chapter 2 turns to the implications for linguistic theory. The success of LLMs challenges the generative (Chomskyan) tradition by showing that complex linguistic behavior can emerge from data-driven learning, without innate rules. Context, frequency, and lexical specificity are central to these models' performance, prompting a reassessment of long-standing debates about competence, universals, and the role of symbolic representation. Yet these systems remain opaque and their relation to human language acquisition remains unresolved.

Chapter 3 addresses how LLMs are reshaping writing itself. Their use in Wikipedia, academia, and everyday communication blurs the line between assistance and authorship. As machine-generated text becomes widespread, questions arise about originality, responsibility, and evaluation. While LLMs increase access and reduce barriers, they also challenge existing norms around trust and intellectual labor. In this new landscape, writing becomes a collaborative, model-mediated act—raising both practical and philosophical concerns.

The second part of the book, titled 'The Risks of Anthropomorphism,' explores how LLMs simulate forms of reasoning, creativity, and moral judgment without possessing consciousness, intentionality, or subjective experience.

Chapter 4 examines the surprising capacity of LLMs to perform tasks associated with reasoning, including analogy, inference, and even elements of theory of mind. Despite lacking a model of the world or internal representations, these systems often succeed in benchmark tests designed for human cognition. Their performance challenges traditional distinctions between symbolic and statistical reasoning and calls for a more nuanced understanding of what counts as 'thinking' when mental states are absent.

In Chapter 5, the focus shifts to creativity. LLMs can now generate plausible poetry, stories, artworks, and code, prompting debate over the nature and authorship of such outputs. Revisiting Margaret Boden's taxonomy of creativity, the chapter argues that, while LLMs fit within combinatorial and exploratory modes, they lack the capacity for intentional transformation or aesthetic evaluation. Yet the integration of their outputs into human creative processes complicates simple distinctions between human and machine creativity.

Chapter 6 turns to moral reasoning and the ethical use of LLMs in normative contexts. From applications like Delphi to evaluations using moral benchmarks, these models increasingly simulate ethical judgment. However, their responses are the result of pattern recognition, not reflection or value commitment. The chapter warns that projecting moral agency onto such systems risks mistaking statistical regularities for normative justification, and overlooks the fundamental absence of responsibility, care, or ethical deliberation in LLMs.

Part III, 'The Social Life of Large Language Models (Their Reach, Roles, and Consequences),' explores how LLMs intervene in social life, public knowledge, and global infrastructures—revealing deep entanglements with inequality, misinformation, and material cost.

Chapter 7 examines how LLMs reproduce social biases and hierarchies through their training data and deployment contexts. Rather than acting as neutral tools, these models reflect dominant cultural norms and reinforce existing power structures, particularly along lines of race, gender, class, and language. The chapter critiques the idea of technical 'fairness' as insufficient, and instead emphasizes the need to understand LLMs as cultural and political artifacts. It argues that bias in LLMs is not simply an error to

be corrected but a reflection of broader social and historical inequalities embedded in the data and institutions that shape these
systems. Addressing such bias, therefore, requires confronting the
deeper politics of representation and power.

Chapter 8 addresses the complex role of LLMs in the spread of
misinformation and disinformation. These systems can generate
convincing but misleading content, blur the boundaries between
fact and fiction, and undermine traditional markers of credibility.
The chapter explores how LLMs destabilize epistemic authority
and reshape the dynamics of public trust, particularly in institutional and media contexts. It argues that confronting this challenge
requires more than technical safeguards—it demands rethinking
how we construct and protect shared frameworks of knowledge
in a digital age.

Chapter 9 turns to the environmental and geopolitical costs of
scaling LLMs. Far from being immaterial, these systems rely on
vast computational infrastructures, extractive resource use, and
global supply chains that carry significant ecological and social
consequences. The chapter situates LLMs within a critique of
technosolutionism, arguing that narratives of innovation often
obscure the harms associated with development at scale. It calls
for evaluating language models not only in terms of performance
or alignment, but also through the lenses of sustainability, global
justice, and planetary responsibility.

Taken together, these chapters offer a philosophical account of
LLMs that is at once technical, critical, and reflective. Rather than
adopting a single disciplinary perspective, the book seeks to map
the diverse conceptual terrain that these models have brought
into focus. Its aim is to provide a structured entry point into the
complex questions raised by the development and deployment
of LLMs, and to offer readers the tools to engage thoughtfully
with technologies that are becoming increasingly embedded in
contemporary life.

At the end of the book, a technical annex provides readers with
a basic introduction to how LLMs work. It is intended as a reference point for those who may wish to better understand the core

mechanisms behind LLMs, and can be consulted at any time to clarify technical concepts mentioned throughout the main chapters. For readers seeking deeper or more specialized knowledge, there is a wide range of detailed and accessible technical blogs available online. We hope this annex proves useful to those who are not fully familiar with the functioning of LLMs and offers a helpful starting point for further exploration.

3 Scope, sources, and methodological considerations

Given the vastness of the intellectual terrain traversed by this book, it is important to acknowledge the necessarily selective nature of its references and thematic scope. The questions raised by LLMs intersect with some of the most enduring and contested areas of philosophical inquiry—most notably, the philosophy of language, the philosophy of mind, and epistemology. These are traditions with long and rich histories, encompassing an immense body of literature. To engage meaningfully with even a fraction of the relevant contributions would require a volume far larger than the one at hand.

At the same time, the recent surge of interest in LLMs has produced an extraordinary volume of contemporary research. Every day, new technical papers, conceptual analyses, ethical commentaries, and speculative reflections are published across disciplines—from computer science and linguistics to cognitive science, media studies, and philosophy. The pace of this intellectual production is remarkable, and no single work could hope to keep pace with the full breadth and depth of the current discourse.

This book is conceived as a short introduction, and its ambition is correspondingly modest. Rather than offering a comprehensive survey, we have aimed to provide a conceptual framework through which readers can begin to make sense of the philosophical implications of LLMs. The references included here are necessarily limited, selected not to be exhaustive but to gesture toward influential debates and foundational texts that can serve as entry points for further exploration. Where possible, we have highlighted major figures and key works in the philosophy of language and mind that resonate with the issues raised by LLMs. Likewise, we draw on a small number of recent studies and discussions from the rapidly evolving field of AI and machine learning research, with the hope that these citations may serve as gateways into the broader landscape.

Ultimately, the goal is not to close the conversation but to open it. Readers are encouraged to pursue their own lines of inquiry, to question the assumptions made here, and to explore the many alternative perspectives that continue to emerge. If this book succeeds in offering some clarity amid the complexity, and in inspiring further engagement with both classical philosophical questions and contemporary technological developments, then it will have fulfilled its intended purpose.

4 Targeted audience

This book is intended for a diverse readership drawn to the philosophical, cultural, and ethical implications of LLMs. While it primarily addresses an academic audience—including scholars in philosophy, linguistics, cognitive science, and the digital humanities—it also welcomes readers from broader intellectual and artistic communities who are interested in how these technologies are reshaping our understanding of language, creativity, and human agency. By engaging with LLMs not merely as technical tools but as cultural artifacts, the book invites reflection on the evolving boundaries between human and machine, the shifting conditions of authorship and interpretation, and the emerging forms of interaction and expression made possible by artificial language.

In addition to serving as a critical reflection on these developments, the book may also function as a textbook for university-level courses concerned with the philosophy of language, artificial intelligence, digital culture, or epistemology. It introduces several well-established problems—such as the nature of meaning, the structure of linguistic understanding, and the epistemic status of machine-generated content—while engaging with a wide range of relevant scholarly sources across disciplines. Throughout, the aim is to offer a balanced and accessible account of the philosophical issues raised by LLMs, clarifying complex debates without oversimplifying them, and proposing interpretations that remain attentive to both conceptual nuance and social consequence.

At the same time, the book does not adopt a position of strict neutrality. It takes a clear stance on several key points: the implications of LLMs for long-standing theories of language and meaning; the risks posed by anthropomorphic framings of these systems, which often obscure the actual mechanisms at work; and the fact that many of the most pressing concerns surrounding LLMs—such as bias, misinformation, and the erosion of trust—are not merely technical challenges but are deeply embedded in broader societal and institutional contexts. The reader is nonetheless encouraged to use the book as a point of departure for their own reflection, including through the critical interrogation of the positions expressed herein. By situating generative AI within a larger philosophical and cultural framework, the book invites not only understanding but also dialogue: an open space for questioning, dissent, and reimagining what it means to live and think with language in an age of artificial authors.

5 Use of generative AI to write this book

The present volume has been composed with the support of generative AI tools, notably LLMs, whose use is both acknowledged and critically examined within the book itself. While the conceptual architecture of the argument, the theoretical positions advanced, and the selection and interpretation of sources are entirely the work of the author, language models have been employed at various stages of the writing process to refine expression, suggest alternative phrasings, or enhance clarity and fluency—particularly in light of the fact that English is not the author's first language.

All factual claims, references, and quotations have been manually verified, and any remaining errors are the sole responsibility of the author. In this sense, the book aims to serve not only as a philosophical inquiry into the implications of generative AI but also as a practical demonstration of how such technologies can be integrated into academic work in a responsible and transparent manner, augmenting human capacities without displacing them.

Reference

McGinn, C. (2017). *Philosophical Provocations.* The MIT Press.

PART I

Sense without reference (large language models and language)

CHAPTER 1

Latent linguistics: the conception of language in large language models

The emergence of large language models (LLMs), such as GPT, Mistral, and Llama, marks a significant moment in the history of language technologies. Initially designed as tools for predicting and generating text, these models have quickly become the foundation for a wide range of applications, including dialogue systems like ChatGPT, code generation tools, writing assistants, search engines, and educational platforms. Their capabilities have sparked intense debate—not only about their practical uses but also about the assumptions they embody regarding language, meaning, and cognition. In many respects, LLMs force a reengagement with long-standing philosophical questions about the nature of language and its relationship to thought and the world.

At the core of LLMs lies a relatively simple objective: Given a sequence of words, predict the most likely continuation. This statistical task, scaled to unprecedented data volumes and computational resources, has produced systems capable of producing

How to cite this book chapter:
Poibeau, T. 2025. *Understanding Conversational AI: Philosophy, Ethics and Social Impact of Large Language Models.* Pp. 13–40. London: Ubiquity Press. DOI: https://doi.org/10.5334/bde.b. License: CC BY-NC 4.0

fluent, coherent, and contextually sensitive text. Yet, despite their apparent sophistication, LLMs operate entirely within the domain of language itself. They do not perceive, act, or refer to the external world. Rather, they model language by detecting patterns of cooccurrence in large textual corpora, generating responses by recombining elements of prior discourse. As such, they instantiate a particular conception of language—one in which meaning arises not from reference or intention but from internal relations among signs.

This vision has important intellectual antecedents. Structuralist and poststructuralist thinkers such as Saussure, Barthes, and Foucault argue that language is not a transparent medium for representing reality but a structured system that shapes thought and subjectivity. They emphasize that meaning is not inherent in words but emerges from differential relations, discursive conventions, and cultural codes. While these were originally philosophical and critical interventions, they now resonate in unexpected ways with the functioning of LLMs—systems that produce language without any stable referent, authorial agency, or grounding in lived experience.

This structuralist affinity, however, is only a starting point. The goal of this chapter is not to situate LLMs within a particular theoretical tradition but to examine the model of language they enact. What kind of representation is involved in a system that learns exclusively from textual data? What semantic capacities, if any, can be attributed to a model that lacks perception, embodiment, or communicative intent? And what do these systems reveal about the limits of language-based modeling more generally?

To explore these questions, we begin with a discussion of structuralist and distributional approaches to meaning, before turning to more recent debates about grounding, reference, and interpretability. The aim is to understand how LLMs manage coherence without understanding, and what their successes and failures can teach us about the nature of linguistic knowledge, the epistemology of statistical modeling, and the boundaries between linguistic form and semantic content.

1 Large language models
and the legacy of structuralist thought

The rise of LLMs not only represents a technological milestone but also brings to life key ideas developed by structuralist and poststructuralist thinkers of the twentieth century. As noted at the introduction of the chapter, these models mirror theories about the nature of language, authorship, and meaning, articulated by figures such as Ferdinand de Saussure, Roland Barthes, and Michel Foucault (and possibly other prominent figures from French theory). Their ideas, once confined to intellectual debates, now find tangible expression in the functioning of LLMs, offering new perspectives on both the capabilities and the limitations of these systems.

One of the central insights from this intellectual tradition is the understanding of language not as a neutral tool manipulated by individuals to express preformed thoughts but as a system that shapes what can be thought and said. Saussure's distinction between *langue* (the collective system of language) and *parole* (individual acts of speech) in a way captures this dynamic.

> Language (fr. *langue*) is a treasure deposited by the practice of speech (fr. *parole*) in the subjects belonging to the same community, a grammatical system existing virtually in each brain, or more precisely, in the brains of a group of individuals; for language is not complete in any one of them, it exists perfectly only in the collective whole. (De Saussure, 2011 [1916])

This passage underscores the fundamentally collective nature of language: It is not an individual possession but a shared system that exists across a community of speakers. Each individual's linguistic competence is necessarily incomplete, relying on the broader structure of *langue* to give meaning and coherence to *parole*. This perspective opens the way to challenging the notion of language as a mere tool for transmitting preexisting thoughts, instead emphasizing how linguistic structures shape and constrain what can be articulated.

From this standpoint, we can draw an initial parallel with LLMs. Like human speakers, these models do not generate language autonomously but operate within a preexisting system built from the linguistic data they have been trained on. They do not 'invent' new rules but assemble responses based on statistical regularities extracted from collective linguistic usage. If we take this idea further, one could argue that each user prompt functions as an instance of *parole*, eliciting a response that reflects the broader *langue* encoded in the model. This process highlights how language, whether in humans or machines, operates within rules and structures that both constrain and shape expression. However, this parallel may offer limited insight into the nature of LLMs or their relationship to language as understood through Saussure's framework.

The structuralist perspective also challenges traditional notions of authorship and originality, ideas further dismantled by Barthes, Foucault, and others in the 1960s. Barthes famously argued for the 'death of the author', suggesting that texts are not the creations of singular, intentional authors but are instead multidimensional spaces where preexisting cultural and linguistic codes intersect.

> A text is made of multiple writings, drawn from many cultures and entering into mutual relations of dialogue, parody, contestation, but there is one place where this multiplicity is focused and that place is not the author, as we have hitherto said it was, but the reader: the reader is the very space in which are inscribed, without any being lost, all the citations a writing consists of. (Barthes, 1977 [1968])

Barthes's argument shifts the focus away from the author as the sole origin of meaning, emphasizing instead the network of texts, references, and cultural codes that shape any piece of writing. Meaning, in this view, is not dictated by a singular authorial intention but emerges through the interplay of existing discourses, activated in the act of reading. The text, rather than being an original, self-contained creation, is a dynamic site of intertextuality, where multiple voices converge and interact.

This perspective finds an uncanny resonance in the way LLMs generate text. Rather than producing language as an expression of individual thought, these models assemble responses by drawing from a vast corpus of preexisting texts. Their outputs, much like the texts described by Barthes, do not originate from a central, authoritative source but are shaped by the interplay of linguistic patterns, cultural references, and statistical associations. In this sense, LLMs not only illustrate but almost literalize Barthes's notion of writing as a palimpsest—an ongoing recombination of citations and influences.

Foucault's critique of the unified subject as the originator of meaning further enriches this understanding. For Foucault, discourse operates independently of individual intent, shaped instead by historical and social rules that determine what can be said and by whom.

> The subject (and its substitutes) must be stripped of its creative role and analysed as a complex and variable function of discourse. (Foucault, 1977 (1969))

Foucault challenges the traditional view of the subject as the origin of meaning, arguing instead that individuals are positioned within discursive structures that shape what can be articulated. Rather than being the source of creative expression, the subject functions within a preexisting system of rules and constraints that define the boundaries of discourse. Meaning, in this framework, is not the product of an autonomous mind but emerges through historically situated practices that govern language and knowledge production.

This perspective offers a productive way to think about how LLMs operate. These models do not generate language as an expression of personal intent but assemble text according to statistical regularities derived from their training data. Their outputs are not the result of a deliberate authorial act but are shaped by the structures embedded in their datasets and training algorithms. Just as Foucault describes discourse as governed by external conditions rather than individual agency, LLMs function within predetermined patterns that dictate how language is produced and reproduced.

The theories of Saussure, Barthes, and Foucault thus provide valuable frameworks for understanding how LLMs function and what they reveal about language itself. Far from being purely abstract, their insights take concrete form in these systems, demonstrating the ongoing relevance of structuralist and poststructuralist thought in the digital age—an idea already explored by various scholars (Underwood, 2023). However, other theoretical approaches may offer a more precise understanding of the conception of language that implicitly emerges from these models. We therefore now turn to a more specific framework: distributional semantics.

2 Large language models through the lens of distributional semantics

Beyond French structuralists, it is a different linguistic and philosophic tradition that we would like to put forward and investigate in relation to LLMs. In particular, we turn to distributional semantics, which posits that linguistic meaning emerges from patterns of cooccurrence in large corpora, a perspective that corresponds closely to the underlying principles of contemporary language models.

2.1 How large language models generate and represent language

LLMs are built to generate text by predicting the next word in a sequence (for more details, see the technical annex at the end of the book). This predictive process relies on the statistical patterns observed in extensive corpora of written language. As the model encounters vast numbers of sentences, it learns which words are most likely to follow others, enabling it to produce text that appears coherent and contextually relevant. When a user begins typing a sentence, the model calculates the probability of possible continuations and selects the most suitable option, repeating this process until the output is complete.

A key component of early language processing systems involved representing words as vectors—numerical arrays that capture patterns of cooccurrence in large text corpora. This process, known as vectorization, does not take into account the specific context in which a word appears. Instead, each word is assigned a single, fixed representation based on its overall distribution across the training data. For example, words like 'apple' and 'orange' often appear in similar contexts (e.g., when discussing fruits), so their vectors will be close in the resulting high-dimensional space, reflecting shared semantic associations. In contrast, words that rarely occur in the same environments will be mapped to more distant positions (see Eisenstein, 2019; Jurafsky and Martin, 2023, for detailed discussions). However, this approach has limitations. For example, it does not distinguish between different meanings of the same word—for instance, 'bank' will have a single embedding whether it refers to a financial institution or the side of a river.

These vector-based representations are important because they enable the mathematical analysis of linguistic relationships. By measuring distances and directions between word vectors, models can estimate how similar certain terms are, and even perform basic analogical reasoning. One of the most frequently cited illustrations of this capacity is the example introduced by Mikolov et al. (2013): the vector equation 'king – man + woman ≈ queen.' This example suggests that word embeddings encode meaningful semantic relationships in geometric form, capturing regularities in how words are used across large corpora.

However, the 'king – man + woman' example has been widely overused and is not as reliable as often claimed. In practice, the operation does not always yield 'queen' as the nearest vector, and its apparent success is partly due to a narrow set of test cases and selective reporting (see Rogers et al., 2017). Moreover, as

Nissim et al. (2020) emphasize, such analogies tend to obscure the limitations of static word embeddings: They ignore context, conflate polysemy, and often reinforce existing stereotypes rather than reveal robust linguistic structure. While these representations do capture certain distributional patterns, they offer only a coarse approximation of meaning and should be interpreted with caution. Their main limitation is that they are static and do not take into account context.

Building on these advances in representation, a major breakthrough came with the introduction of the transformer architecture (Vaswani et al., 2017). Unlike earlier models that processed text sequentially, word by word, the transformer considers all words in a sentence—or even a longer passage—simultaneously. This is made possible by a mechanism called self-attention, which allows the model to assess the relationships between all words in the input. Instead of focusing on individual tokens or narrow contexts, the model dynamically adjusts its attention to capture the most relevant information across the entire input.

The model can thus assign different meanings to the same word depending on its context. This means that the same word can have different internal representations depending on how it is used in a sentence. As previously mentioned with the word 'bank,' such models can distinguish between meanings that static embeddings would collapse into a single representation. This ability to adapt word meaning based on context is especially valuable for addressing polysemy, where a single surface form corresponds to multiple semantic interpretations.

These rich, flexible representations also play a key role in text generation. As the model predicts each new word in a sentence, it draws not only on the immediate word before it but on a nuanced understanding of the full context it has already processed. This allows the model to maintain coherence, handle long-range dependencies, and choose words that make sense not just syntactically but semantically and thematically as well. The transformer architecture is thus a central reason why modern LLMs can produce text that feels fluid, appropriate, and often surprisingly human-like. Its greatest advantage lies in its ability to be efficiently parallelized, which makes large-scale training feasible.

In sum, LLMs learn to predict subsequent words by analyzing text at a massive scale and encoding language elements as vectors. These representations make it possible to transform raw text into a mathematical space where semantic relationships can be identified and leveraged. Although LLMs do not possess genuine understanding in the human sense, the manner in which they represent and combine words has proven powerful enough to generate text that is both contextually coherent and suggestive of deeper linguistic structures.

2.2 The distributional hypothesis and its relevance to large language models

Modern LLMs rely on the principle that words and phrases derive part of their meaning from the contexts in which they frequently appear. By examining these contextual cooccurrences in massive text corpora, LLMs learn numerical representations that capture the relationships among linguistic elements. This approach hinges on a deeper theoretical framework often referred to as distributional semantics, which posits that meaning can be modeled through the statistical patterns of word usage.

This framework traces its origins to the mid-twentieth century, when linguists, mathematicians, and philosophers first explored how language meaning is tied to the contexts in which words occur. Influential figures such as John Rupert Firth (a linguist), Zellig Harris (a linguist and mathematician), and Ludwig Wittgenstein (a philosopher) laid the foundations of distributional semantics by emphasizing the contextual and relational dimensions of language. Around the 1950s, their insights paved the way for computational approaches that harness statistical patterns to model meaning. J. R. Firth's famous assertion that 'you shall know a word by the company it keeps' (1957) captures the essence of distributional semantics. Firth argued that the meaning of a word is closely tied to its habitual cooccurrences with other words in linguistic contexts. This perspective underpins the idea that words appearing in similar contexts tend to share semantic properties. Zellig Harris expanded on this concept, introducing the distributional hypothesis, which states that linguistic elements occurring in similar environments tend to have related meanings (Harris, 1954).

> If we consider words or morphemes A and B to be more different in meaning than A and C, then we will often find that the distributions of A and B are more different than the distributions of A and C. In other words, difference of meaning correlates with difference of distribution. (Harris, 1954)

Both Firth and Harris shifted the focus of linguistic inquiry from introspection to the observable patterns of language use, emphasizing the importance of data-driven analysis.

Wittgenstein's philosophy is also frequently cited by papers dealing with distributional semantics. In his *Philosophical Investigations* (1953), the philosopher suggests that the meaning of a word is not fixed but emerges through its use in language games, or patterns of linguistic behavior within specific social contexts.

> The meaning of a word is its use in the language. (Wittgenstein, 2009 (1953), §43)

> We can also think of the whole process of using words ... as one of those games by means of which children learn their native language. I will call these games 'language-games' and will sometimes speak of a primitive language as a language-game. (Wittgenstein, 2009 (1953), §7)

Through the idea of 'language games,' the philosopher emphasizes that the meaning of a word emerges from how it is used within particular social and contextual practices. Each 'game' has its own rules, participants, and aims, and language functions differently depending on which game is being played. This emphasis on use matches with the distributional view, where meaning is derived from the statistical associations of words in corpora, reflecting their functional roles rather than intrinsic properties.

The practical implementation of these ideas in natural language processing (NLP) relies on vectorization techniques, where words or linguistic elements are represented as points in a high-dimensional vector space. This approach operationalizes the insights of Firth, Harris, and Wittgenstein by quantifying the 'company' that words keep. For example, word embeddings such as Word2Vec (Mikolov et al., 2013) or GloVe (Pennington et al., 2014) compute dense vector representations of words based on their cooccurrence patterns in large corpora. These embeddings encode semantic relationships, allowing words with similar contexts to cluster closely in the vector space. Such methods reflect the theoretical foundation laid by distributional semantics, translating abstract linguistic principles into computational algorithms.

2.3 Text-based meaning, or the nonreferential nature of large language model representations

The representations that LLMs manipulate are fundamentally rooted in statistical patterns of word cooccurrence. During training, these models ingest billions of tokens, learning which words and phrases tend to appear together in context. From this process, they adjust billions of parameters, including initial word embeddings and the weights of complex transformer layers. These parameters allow the model to build high-dimensional vector representations that are not static but dynamically refined through multiple layers of self-attention and feedforward transformations. As a result, modern LLMs capture much more than simple cooccurrence; they encode nuanced, contextualized information about text (and, indirectly, about syntax, semantics, and discourse structure).

This training takes place without any direct reference to external, real-world objects or situations: The only input is text, and the only criterion for success is learning to predict the next token with increasing accuracy. As a result, the semantic 'atoms' or components that emerge in LLM embeddings are inferred purely from linguistic regularities, rather than from explicit conceptual or ontological frameworks.

While this text-driven approach aligns with distributional semantics, it also resonates with philosophical accounts that emphasize meaning as use, notably Wittgenstein's notion of language games, as seen in the previous section. Although LLMs do not participate in real-world practices, their exposure to diverse genres—ranging from legal documents to social media—allows them to approximate some forms of context-sensitivity. For instance, the word 'pitch' may appear in contexts related to music, sports, or business, and through repeated exposure to such patterns, LLMs learn to produce coherent continuations across domains.

However, the underlying processes differ significantly from what Wittgenstein described. Wittgenstein's language games are deeply embedded in what he called 'forms of life,' encompassing shared cultural, physical, and social contexts. An LLM has no direct access to these embodied experiences; it infers context indirectly from patterns of language usage alone. While it can synthesize context clues from vast textual data, it does not participate in, or observe, real-world interactions and thus does not learn language in the same way humans do when they 'play' actual language games in everyday life. In that sense, while LLMs echo certain aspects of Wittgenstein's idea—mainly that meaning depends on context—they do so in an abstract, text-only manner. They reflect how words are used in written communications but cannot grasp the nonlinguistic, practical dimensions of human language games. This could be nuanced by taking into account the role of human feedback during supervised fine-tuning, as in ChatGPT, which introduces social evaluation and correction of the model's outputs, though still without any embodied participation.

Another major feature of LLMs is their ability to represent meaning through latent dimensions in a high-dimensional vector space. This might recall theoretical frameworks that posit a hypothesized universal inventory of semantic primitives, which function as building blocks for constructing more complex meanings. For example, Katz and Fodor's componential theory (1963) proposes that word meanings could be analyzed in terms of a finite set of universal semantic features. Similarly, Anna Wierzbicka's theory of 'semantic primes' (1996) suggests that all complex meanings can be constructed from a small set of irreducible, language-independent semantic elements. These proposals represent attempts to explicitly define a universal set of minimal components of meaning. In contrast, LLMs do not start from any such predefined set. Instead, their latent representations emerge from statistical patterns in large-scale text data. If they resemble features akin to semantic primes, this is purely a byproduct of data-driven learning rather than a built-in theoretical commitment.

In practice, one may find partial correspondences between certain latent dimensions and recognized semantic features. By examining how words cluster or how analogy-like operations (e.g., 'king – man + woman ≈ queen'; see above Section 2.1) play out in the embedding space, it may be possible to sometimes interpret the vectors as reflecting meaning-related properties. However, there is no explicit mechanism that guarantees these dimensions will map neatly onto a particular theoretical framework of conceptual primitives. Each dimension emerges from the interplay of countless linguistic contexts during training, rather than from a carefully specified semantic architecture.

The success of LLMs suggests that learning solely from text can produce models with robust linguistic capacities. The sheer volume of textual data available, spanning a wide variety of topics, provides an enormous reservoir of implicitly encoded human knowledge. Consequently, LLMs have been deployed for a range of tasks beyond mere word prediction—translation, summarization, question-answering, poetry generation, and more. Even though they are only trained to predict words in sequences, their ability to handle such diverse tasks has often exceeded initial expectations. Because the training corpora include texts reflecting many facets of human culture, science, and everyday communication, the models can recombine these textual fragments in ways that produce coherent, contextually relevant, and often highly creative outputs.

Despite this, a vocal minority within the NLP community has challenged this interpretation, or at least questioned whether these models genuinely encode semantics. This critique reflects ongoing debates about the nature of meaning in language and the extent to which LLMs truly understand or merely approximate semantic relationships.

3 Stochastic parrots, and the real nature of meaning

The rapid advancement of LLMs has reignited long-standing debates about the nature of linguistic meaning and the limits of computational approaches to language. Central to this discussion is the question of whether statistical associations alone can constitute meaning or whether genuine comprehension requires something more—such as embodiment, intentionality, or interaction with the external world.

One of the most influential critiques of LLMs in this regard comes from Bender et al. (2021), who argue that these systems function as 'stochastic parrots'—sophisticated pattern matchers that generate plausible text without any true grasp of what they are saying.

3.1 The stochastic parrot hypothesis: Bender et al. on meaning and understanding

Bender et al. (2021) characterize LLMs as 'stochastic parrots,' highlighting that these systems generate text by probabilistically assembling linguistic elements rather than by leveraging any true understanding. They base this critique on the observation that meaning in natural language is not merely a matter of form or syntax but requires an external grounding—an authentic link between the words and the entities or concepts they denote. In other words, LLMs may produce coherent, contextually appropriate responses but the underlying processes are geared toward pattern replication rather than genuine reference or intentionality. As a result, while LLMs can simulate human-like conversation, they do not capture the essence of what it means to understand language (Marcus and Davis, 2020).

A closely related argument appears in an article by Bender and Koller (2020), who explore how LLMs fail to capture the richness of semantic meaning. Their paper directly challenges the assumption that increasingly large datasets and more powerful models will inevitably yield systems that comprehend language as humans do. Instead, the authors claim that genuine natural language understanding depends on grounding linguistic expressions in real-world experience or perception. Merely learning statistical correlations among tokens does not endow a model with the capacity to connect utterances to real objects, events, and contexts. Bender and Koller argue that true meaning arises at the interface of linguistic form, cognitive processes, and situated interactions— an interface that today's LLMs have yet to meaningfully access.

The claim that LLMs are not grounded in external reality, which might undermine their ability to understand language, relates to long-standing philosophical debates about the nature of meaning. Frege (1993 [1892]) argues that linguistic expressions must have a reference (*Bedeutung*) alongside sense (*Sinn*), suggesting that semantics requires a link to real-world entities. Putnam (1975) developed this view further by showing that 'meaning ain't just in the head,' implying that context and environment play a crucial role in securing reference. These positions are broadly referentialist, seeing meaning as partly dependent on the capacity of words to hook onto things in the world.

Similarly, Searle's (1980) famous Chinese room argument describes a scenario in which a person who does not speak Chinese can nevertheless produce appropriate responses by following syntactic rules. This illustrates a situation where symbols are manipulated purely syntactically, without any true referential grounding. Although the system can generate responses that might fool observers, it lacks any genuine understanding of what the symbols stand for. This reflects an intuition aligned with direct-reference theory, in the sense that meaning arises from the capacity to link language to the surrounding environment, rather than merely from manipulating patterns. Bender et al.'s (2021) critique of LLMs (and arguably also Bender and Koller, 2020) can be read in this tradition—although this parallel is our own interpretation, as these references are not explicitly cited in Bender et al.'s article. If a machine merely shuffles linguistic forms without anchoring them to external referents, it cannot properly be said to 'understand' language.

Even prior to the emergence of LLMs, scholars proposed several avenues for addressing the grounding problem, most notably through multimodal systems. Harnad's (1990) 'symbol grounding problem' set the stage for investigating how to link abstract linguistic symbols to sensory experiences of the world. More recent proposals, such as that of Bisk et al. (2020), argue that real-world grounding might be achieved by coupling linguistic models with visual, auditory, or even sensorimotor data, thereby connecting text to perception and action. While such approaches may move AI systems closer to referential semantics, critics maintain that merely integrating multiple data streams does not necessarily result in genuine understanding. True semantics, in this view, arises from the capacity to intentionally refer to objects and experiences. This would require an internal conceptual framework that interprets and reinterprets language in light of ongoing perception and interaction.

For Bender et al., the gap between statistical fluency and genuine reference underscores an essential difference between human cognition and current AI systems. Indeed, humans acquire language through embodied, situated experiences: Children learn words by associating sounds and contexts with meaningful aspects of lived experience. By contrast, LLMs passively absorb massive corpora of text without firsthand encounters of the phenomena described, which arguably leaves them blind to the external world. While some researchers, such as Lake et al. (2017), emphasize building AI that learns and thinks more like people—thereby deeply integrating perception, cognition, and action—Bender et al. caution that, as long as we rely exclusively on text-based training, we risk inflating the capabilities of LLMs beyond what they can truly deliver.

In sum, Bender et al. (2021) and Bender and Koller (2020) challenge the notion that scaling up language models automatically yields meaningful understanding. They argue that the capacity to produce plausible-sounding text by 'stochastic' means should not be mistaken for semantic competence. Following the tradition of Frege, Putnam, and Searle, they maintain that language must be grounded in external reality to be truly meaningful. The result is a strong call for reevaluating the true nature of semantics, prompting researchers to seek AI architectures that integrate linguistic form with perceptual, cognitive, and interactive grounding—so that AI systems might one day progress beyond merely mimicking understanding and toward exhibiting genuine, referential language use.

3.2 The limits of ungrounded language: hallucination and synthetic plausibility

One of the most significant limitations of LLMs lies in their inability to guarantee the truthfulness of the text they generate—a direct consequence of their lack of grounding in the external world. While these models excel at producing coherent, contextually appropriate, and grammatically correct language, they do so by predicting the most likely sequence of tokens based on patterns learned from their training data. They are, in essence, sophisticated pattern completion machines, not epistemic agents capable of verifying the correspondence between their outputs and any external reality.

To understand the link with the preceding discussion, it is helpful to distinguish between two kinds of grounding (Mollo and Millière, 2023, even distinguish five types of grounding). On the one hand, *referential grounding* concerns the fundamental capacity of words or symbols to stand for entities, properties, or events in the external world—a link traditionally theorized in philosophy of language as a condition of meaningful reference, as seen in the previous section. On the other hand, *factual grounding* refers to whether statements built from those symbols accurately describe real-world facts. Although these dimensions are conceptually distinct, they are deeply connected: Factual grounding presupposes referential grounding. If a symbol lacks a genuine connection to what it purports to represent, then any statements constructed from it cannot be reliably assessed for truth, since truth conditions depend on stable referential ties.

A particularly salient manifestation of this limitation is the phenomenon commonly referred to as hallucination. In the context of LLMs, hallucination designates the generation of information that is factually incorrect, fabricated, or unsupported by any real-world source (Ji et al., 2023). Unlike human errors, which may result from memory lapses or misunderstanding, hallucinations in LLMs are emergent artifacts of their statistical architecture. Because the model by itself lacks referential grounding—that is, it has no genuine connection to the things its symbols purport to describe—it cannot access or verify the truth conditions of its outputs. It merely constructs text that is locally plausible within its training distribution, without any epistemic guarantee of factual accuracy.

To conceptualize this, we can introduce the notion of 'synthetic plausibility.' Borrowed from biology, where synthetic plausibility describes artificially generated phenomena that mimic natural forms without reproducing their underlying mechanisms, the term aptly characterizes how LLMs operate. They generate language that appears plausible—syntactically fluent, semantically coherent, and often stylistically convincing—without any guarantee of factual accuracy. Their outputs are synthetic constructs shaped by exposure to vast textual data, but devoid of any intrinsic connection to verifiable states of affairs.

In most cases, because LLMs are trained on extensive datasets sourced from human-authored text, the information they reproduce corresponds to generally accurate or commonly accepted knowledge. However, this reliance on textual patterns alone makes it difficult to control or anticipate when and why deviations occur. It is precisely in such moments that the absence of grounding reveals itself starkly. Classic examples from early deployments of LLM-based systems include nonsensical or fabricated outputs such as the generation of nonexistent biological entities ('horse's eggs'), invented scientific references, or apocryphal quotations attributed to well-known figures. These instances expose the model's lack of access to an external fact-checking mechanism.

While advancements in model architecture, fine-tuning techniques, and postprocessing methods have significantly reduced the frequency and prominence of such errors, hallucinations remain possible—and indeed inevitable—given the underlying limitations (and generative nature) of the technology. As prominent researchers such as Yann LeCun have frequently pointed out,[1] no matter how large or refined these models become, they fundamentally lack an intrinsic mechanism to validate the truth of their outputs. The issue is not merely technical but epistemic: Without explicit grounding, there is no bridge between linguistic form and factual content beyond what is implicitly encoded in the training data.

This tension—between the surface-level coherence of generated language and the absence of guaranteed factual grounding—constitutes one of the most fundamental constraints on the epistemic reliability of LLMs. It invites ongoing philosophical reflection not only on the nature of meaning and understanding in artificial systems but also on the broader societal implications of deploying technologies whose outputs, while syntactically convincing, may drift unpredictably from truth.

---

[1] In this 2024 YouTube video for example: Yann LeCun: Limits of LLMs | Lex Fridman Podcast Clips: https://www.youtube.com/watch?v=1lHFUR-yD6I.

4 Communication with (limited) grounding, a pragmatic perspective

The absence of metaphysical grounding does not necessarily preclude LLMs from achieving highly functional forms of communication. In practice, systems like ChatGPT illustrate that effective dialogue does not always depend on deep semantic understanding. Conversation, after all, is not reducible to reference and truth conditions alone; it also involves pragmatics, pattern recognition, and forms of social interaction that these models approximate with considerable success. The widespread adoption of LLMs for text drafting, question-answering, and creative writing suggests that many valuable applications do not strictly require fully grounded semantics.

4.1 Beyond reference: inferential and pragmatic accounts of meaning

While Bender et al.'s position is widely endorsed, alternative frameworks offer different insights into how LLMs might achieve a kind of meaningfulness or grounding, at least in a derivative sense. One promising approach is inferentialism, most notably advanced by Brandom (2000). Inferential semantics contends that the meaning of an expression is determined not by a direct causal link to external referents but rather by its role within a network of inferences—what conclusions it permits and what reasons would justify its use. Applied to LLMs, this perspective invites us to consider how their outputs participate in a 'space of reasons,' shaped by patterns of entailment, contradiction, and coherence acquired through exposure to human linguistic practices. Even if their symbols lack metaphysical, referential grounding, LLMs may be described as inferentially grounded in the sense that their utterances fit coherently within the rule-governed discursive practices of human communication. In this view, LLMs display a form of derived pragmatic competence, because their language is interpreted and evaluated by human interlocutors within shared normative frameworks, rendering their outputs intelligible and actionable despite the absence of intrinsic reference-fixing capacities.

While Brandom's inferentialism offers a valuable framework for thinking about meaning, it is a general philosophical theory not specifically designed for computational models. By contrast, a recent and explicitly LLM-focused proposal by Mollo and Millière (2023) examines how language models might achieve forms of referential grounding. They argue that fine-tuning can guide models toward internal states that track real-world entities and events, by incorporating human preferences tied to extra-linguistic standards such as factual accuracy. They also suggest that even pretraining alone may, in certain cases, favor internal representations with world-involving content. Since predicting the next token often requires internalizing the underlying structure of the data, LLMs may develop representations that reflect aspects of the world. In this view, referential grounding emerges not from direct interaction with reality but from the statistical and causal patterns embedded in the training data.

These alternative frameworks remind us that grounding, in its various senses, is a complex notion. Human communication itself frequently operates without strict referential grounding: Ordinary conversation relies on heuristics, shared conventions, and pragmatic inference rather than precise, verifiable links to external reality (Clark, 1996). Language games, rhetorical devices, and even routine misunderstandings illustrate how linguistic interaction can remain meaningful despite the absence of rigid semantic anchoring. From this vantage point, LLMs underscore an important distinction: While robust grounding is indispensable for certain applications—such as scientific reasoning, legal argumentation, or evidence-based policymaking—it is not a prerequisite for every communicative context.

Ultimately, the critique offered by Bender et al. (2021) rightly highlights some significant limitations of contemporary language models but it does not nullify their practical utility. Rather, it invites deeper reflection on what counts as meaningful language use, and whether different domains require different forms of understanding. Although today's LLMs do not achieve true reference or intentionality in the human sense, their capacity to generate coherent and pragmatically useful text suggests that language can remain functional even in the absence of traditional forms of grounding.

4.2 Toward partial grounding: multimodality, reinforcement learning, and hybrid architectures

Several recent developments have sought to bridge the gap between linguistic form and external reality—introducing elements of factual grounding, and in some cases even referential grounding (see Mollo and Millière, 2023, for an extended discussion). Among these, the integration of multimodality is often presented as a significant advance (see Section 3.1). Multimodal models, which combine textual data with images or videos, expose language models to nonlinguistic forms of information, aligning textual descriptions with visual representations (Alayrac et al., 2022). By learning associations between words and images, such systems seem to broaden their representational scope, acquiring an apparent 'understanding' of the world.

However, this incorporation of multimodal data does not fundamentally alter the grounding problem. The model learns to correlate text and visual patterns but this remains a process of statistical association rather than genuine perception. The model has no access to the truthfulness of the visual content or its relation to the external world. It does not perceive images as intentional agents do, nor does it possess the capacity to verify or question the accuracy of its inputs. As such, multimodality enhances representational richness without resolving the epistemic gap between symbol and referent.

A more structurally significant technique for introducing a form of partial grounding is reinforcement learning with human feedback (RLHF) (Ouyang et al., 2022). RLHF is now widely adopted in the development of chatbot systems based on LLMs, including prominently in ChatGPT and similar applications. It plays a central role in the 'chat' aspect of these systems. The process involves fine-tuning the model by presenting multiple candidate outputs and having human annotators rank them according to various criteria: factual accuracy, relevance, coherence, politeness, or ethical soundness. Over successive iterations, the model learns to prefer responses that align more closely with human preferences and communicative norms.

What makes RLHF particularly notable is that it introduces an external evaluative loop into the model's training process. Human feedback becomes a mechanism for selecting not only factually correct answers but also those deemed most appropriate or useful in specific conversational contexts. Unlike training purely on large, static datasets, RLHF allows for dynamic adjustment based on human judgment, embedding social and normative expectations directly into the model's behavior.

While RLHF does not, strictly speaking, provide direct epistemic access to the world—it relies on human assessments, themselves fallible and culturally situated—it establishes an intersubjective form of validation. The model is guided not solely by statistical likelihoods but by explicit human evaluation. Philosophically, this approach echoes elements of pragmatic theories of meaning, where meaning emerges through use, correction, and shared practices, rather than through static reference alone. RLHF does not eliminate the absence of grounding but it mitigates it by embedding the model in an ongoing, socially mediated evaluative process.

Beyond RLHF, additional strategies have emerged to improve the factual reliability of language models. A widely adopted method involves coupling LLMs with structured knowledge sources, such as curated knowledge bases or knowledge graphs. These resources encode verified information about entities, relationships, and facts, allowing models to ground their generative output in trustworthy data. This hybrid approach is increasingly integrated into modern systems. For example, Google's Gemini leverages the Google Knowledge Graph to enhance its answers with authoritative references, while Amazon's Alexa combines LLM capabilities with its internal structured data repositories and skill-based APIs to deliver accurate and context-aware responses. Such integrations reduce the risk of hallucinations by providing external, verifiable reference points—an architecture now central to many search and assistant platforms.

Another widely discussed approach is retrieval-augmented generation (RAG, Lewis et al., 2020; Fan et al., 2024). Here, the model is paired with a retrieval system capable of sourcing documents or factual material in real time, dynamically informing its responses. RAG systems allow models to reference up-to-date knowledge beyond the temporal limitations of their training data. However, the efficacy of RAG depends crucially on the quality of the retrieval sources. If the system draws indiscriminately from the open internet, the problem of unverifiable or erroneous information persists. Reliable grounding (at least, factual grounding) requires careful curation of retrieval sources, emphasizing the limits of relying solely on retrieval for epistemic soundness.

Taken together, these strategies suggest that grounding in LLMs is best understood not as a binary property but as a layered, procedural achievement. Multimodal data, RLHF, knowledge bases, and retrieval systems all contribute to enhancing the model's functional alignment with human expectations of truthfulness, relevance, and coherence. Philosophically, this points to a relational conception of meaning, in line with thinkers like Wittgenstein (1953; see also Section 2.3 above), or more recently Davidson (1984) and Brandom (1994), where understanding is situated within social practices, inferential roles, and communal validation processes.

While none of these techniques fully resolves the foundational issue of grounding, they establish partial mechanisms that embed language models within broader ecosystems of human judgment, structured knowledge, and external information sources. In doing so, they mark important steps toward making AI-generated language more reliable, contextually appropriate, and pragmatically useful—even in the absence of full experiential anchoring.

References

Alayrac, J.-B., Donahue, J., Luc, P., Miech, A., Barr, I., Hasson, Y., Lenc, K., Mensch, A., Millican, K., Reynolds, M., Ring, R., Rutherford, E., Cabi, S., Han, T., Gong, Z., Samangooei, S., Monteiro, M., Menick, J., Borgeaud, S. … Simonyan, K. (2022). Flamingo: A visual language model for few-shot learning. In *Proceedings of the 36th International Conference on Neural Information Processing Systems (NeurIPS '22), Red Hook, NY, USA* (pp. 23716–23736). Curran. https://dl.acm.org/doi/10.5555/3600270.3601993

Barthes, R. [1977 (1968)]. The death of the author. In S. Heath (Ed. and Trans.), *Image, music, text* (pp. 142–148). Fontana Press.

Bender E. M., and Koller, A. (2020). Climbing towards NLU: On meaning, form, and understanding in the age of data. In *Proceedings of the 58th Annual Meeting of the Association for Computational Linguistics* (pp. 5185–5198). Association for Computational Linguistics. https://aclanthology.org/2020.acl-main.463/

Bender, E.M., Gebru, T., McMillan-Major, A., and Shmitchell. S. (2021). On the dangers of stochastic parrots: Can language models be too big? 🦜. In *Proceedings of the 2021 ACM Conference on Fairness, Accountability, and Transparency (FAccT '21), New York, NY, USA* (pp. 610–623). Association for Computing Machinery. https://doi.org/10.1145/3442188.3445922

Bisk, Y., Holtzman, A., Thomason, J., Andreas, J., Bengio, Y., Chai, J., Lapata, M., Lazaridou, A., May, J., Nisnevich, A., Pinto, N., and Turian, J. (2020). Experience grounds language. In *Proceedings of the 2020 Conference on Empirical Methods in Natural Language Processing (EMNLP)* (pp. 8718–8735). Association for Computational Linguistics. https://aclanthology.org/2020.emnlp-main.703/

Brandom, R. B. (1994). *Making it explicit. Reasoning, representing, and discursive commitment*. Harvard University Press.

Brandom, R. B. (2000). *Articulating reasons: An introduction to inferentialism*. Harvard University Press.

Clark, H. H. (1996). *Using language*. Cambridge University Press.

Davidson, D. (1984). *Inquiries into truth and interpretation*. Oxford University Press.

De Saussure, F. (2011 [1916]). *Course in general linguistics*. Columbia University Press.

Eisenstein, J. (2019). *Introduction to natural language processing.* The MIT Press.

Fan, W., Ding, Y., Ning, L., Wang, S., Li, H., Yin, D., Chua, T., and Li, Q. (2024). A survey on RAG meeting LLMs: Towards retrieval-augmented large language models. In *Proceedings of the 30th ACM SIGKDD Conference on Knowledge Discovery and Data Mining (KDD '24), New York, NY, USA* (pp. 6491–6501). Association for Computing Machinery. https://doi.org/10.1145/3637528.3671470

Firth, J. R. (1957). *Papers in linguistics, 1934–1951.* Oxford University Press.

Foucault, M. [1977 (1969)]. What is an author? In D. F. Bouchard and S. Simon (Trans. and Eds.), *Language, counter-memory, practice* (pp. 113–138). Cornell University Press.

Frege G. [1993 (1892)]. On sense and reference. In A. W. Moore (Ed.), *Meaning and reference* (pp. 23–42). Oxford University Press.

Harnad, S. (1990). The symbol grounding problem. *Physica D: Non-linear Phenomena, 42*(1–3), 335–346. https://doi.org/10.1016/0167-2789(90)90087-6

Harris, Z. (1954). Distributional structure. *Word, 10*(2–3), 146–162.

Ji, Z., Lee, N., Frieske, R., Yu, T., Su, D., Xu, Y., Ishii, E., Jin Bang, Y., Madotto, A., and Fung, P. (2023). Survey of hallucination in natural language generation. *ACM Computing Surveys, 55*(12), article 248 (December 2023). https://doi.org/10.1145/3571730

Jurafsky, D., and Martin, J. H. (2023). *Speech and language processing* (3rd ed.). https://web.stanford.edu/~jurafsky/slp3/

Katz, J. J. and Fodor, J. A. (1963). The structure of a semantic theory. *Language, 39*(2), pp. 170–210. https://wwwhomes.uni-bielefeld.de/mkracht/kurse/ws2017-18/merkmale/KatzFodor_Structure_of_Semantic_Theory.pdf

Lake, B. M., Ullman, T. D., Tenenbaum, J. B., and Gershman, S. J. (2017). Building machines that learn and think like people. *Behavioral and Brain Sciences, 40*, e253. https://doi.org/10.1017/S0140525X16001837

Lewis, P., Perez, E., Piktus, A., Petroni, F., Karpukhin, V., Goyal, N., Küttler, H., Lewis, M., Yih, W., Rocktäschel, T., Riedel, S. and Kiela, D. (2020). Retrieval-augmented generation for knowledge-intensive NLP tasks. In *Proceedings of the 34th International Conference on Neural Information Processing Systems (NeurIPS '20)* (pp. 9459–9474). Article 793, Curran Associates Inc. https://dl.acm.org/doi/abs/10.5555/3495724.3496517

Marcus, G., and Davis, E. (2020). GPT-3, bloviator: OpenAI's language generator has no idea what it's talking about. *MIT Technology Review*, August 28, 2020. https://www.technologyreview.com/2020/08/22/1007539/gpt3-openai-language-generator-artificial-intelligence-ai-opinion/

Mikolov, T., Chen, K., Corrado, G., and Dean, J. (2013). Efficient estimation of word representations in vector space. *arXiv:1301.3781*. https://arxiv.org/abs/1301.3781

Mollo, G., and Millière, R. (2023). The vector grounding problem. arXiv preprint. *arXiv:2304.01481*. https://arxiv.org/abs/2304.01481

Nissim, M., van Noord, R. and van der Goot, R. (2020). Fair Is better than sensational: Man is to doctor as woman Is to doctor. *Computational Linguistics*, MIT Press, 46(2):487–497. https://aclanthology.org/2020.cl-2.7

Ouyang, L., Wu, J., Jiang, X., Almeida, D., Wainwright, C. L., Mishkin, P., Zhang, C., Agarwal, S., Slama, K., Ray, A., Schulman, J., Hilton, J., Kelton, F., Miller, L., Simens, M., Askell, A., Welinder, P., Christiano, P., Leike, J., and Lowe, R. (2022). Training language models to follow instructions with human feedback. In *Proceedings of the 36th International Conference on Neural Information Processing Systems (NIPS '22), Red Hook, NY, USA* (pp. 27730–27744). Curran. https://arxiv.org/abs/2203.02155

Pennington, J., Socher, R. and Manning. C. (2014). GloVe: Global vectors for word representation. In *Proceedings of the 2014 Conference on Empirical Methods in Natural Language Processing (EMNLP)* Doha, Qatar (pp. 1532–1543). Association for Computational Linguistics. https://aclanthology.org/D14-1162/

Putnam, H. (1975). Language, mind, and knowledge. *Minnesota Studies in the Philosophy of Science, 7*, 131–193. Retrieved from the University Digital Conservancy, https://hdl.handle.net/11299/185225

Rogers, A., Drozd, A. and Li, B. (2017). The (too many) problems of analogical reasoning with word vectors. In *Proceedings of the 6th Joint Conference on Lexical and Computational Semantics (*SEM 2017)* Vancouver, Canada (pp. 135–148). Association for Computational Linguistics. https://aclanthology.org/S17-1017/

Searle, J. (1980). Minds, brains and programs. *Behavioral and Brain Sciences, 3*(3), 417–457. https://doi.org/10.1017/S0140525X00005756

Underwood, T. (2023, June 29). The empirical triumph of theory. *Critical Inquiry* [Blog]. https://critinq.wordpress.com/2023/06/29/the-empirical-triumph-of-theory/

Vaswani, A., Shazeer, N., Parmar, N., Uszkoreit, J., Jones, L., Gomez, A. N., Kaiser, Ł., and Polosukhin, I. (2017). Attention is all you need. *Advances in Neural Information Processing Systems (NeurIPS)* Long Beach, *30*. https://arxiv.org/abs/1706.03762

Wierzbicka, A. (1996). *Semantics: Primes and universals.* Oxford University Press.

Wittgenstein, L. [2009 (1953)]. *Philosophical investigations.* Wiley-Blackwell.

CHAPTER 2

From the design of large language models to a reassessment of linguistic theory

It is often observed that a significant gap separates the practical development of language models (examined in the previous chapter) from the questions and methods of linguistic theory, as well as from broader philosophical approaches to language. Despite this divide, the remarkable success of large language models (LLMs) and the rapid progress of computational language modeling may offer valuable insights for linguistic theory and the description of language itself. Although LLMs do not provide a direct account of how language is processed in the human brain, their mechanisms and behaviors can nonetheless illuminate certain properties of language as a system, prompting a reconsideration of long-standing theoretical assumptions.

A central starting point for this discussion is the notion of context, which has proven fundamental to explaining the performance of modern language models. In particular, the ability of words—and linguistic units more generally—to adjust their meaning flexibly depending on surrounding linguistic material appears essential to their success. Yet context, despite its centrality in

How to cite this book chapter:
Poibeau, T. 2025. *Understanding Conversational AI: Philosophy, Ethics and Social Impact of Large Language Models.* Pp. 41–63. London: Ubiquity Press. DOI: https://doi.org/10.5334/bde.c. License: CC BY-NC 4.0

linguistic discourse, has historically remained somewhat vague and under-theorized, with no fully formalized definition across frameworks. It is precisely the sophisticated handling of contextual dependencies in today's LLMs that underlies much of their power, and that calls for renewed attention in linguistic theorizing.

Building on this observation, several researchers—most notably Steven Piantadosi, along with others working in usage-based, emergentist, or constructionist traditions—have argued that the empirical success of LLMs challenges the assumptions of the Chomskyan paradigm. In particular, these scholars question whether innate, domain-specific structures are as central to linguistic competence as generative grammar has traditionally proposed. Instead, they argue that the rich statistical patterns captured by LLMs point to a different conception of language—one shaped by patterns of contextual distribution rather than by hardwired universal rules, even if these patterns do not directly reflect human experience. This perspective explicitly confronts the poverty-of-the-stimulus argument and reopens debates about what linguistic knowledge is and how it is acquired.

In this chapter, we explore these debates in depth. We begin by examining how LLMs implicitly model key linguistic phenomena, particularly the influence of context, frequency effects, and lexical specificity. We then revisit the historical debates that shaped linguistic theory in the twentieth century—including the generative/semantic controversies—showing why these histories remain relevant to current discussions. Finally, we consider alternative frameworks that may be more compatible with, and potentially enriched by, the empirical findings emerging from modern large-scale language models. In doing so, this chapter aims to show that LLMs do not simply pose engineering challenges but rather open up a broader reassessment of the goals, concepts, and methods of linguistic theory itself.

1 Context is all you need

One of the most pressing questions in natural language processing (NLP) today is why LLMs exhibit such remarkable proficiency at generating coherent, contextually rich text compared to earlier systems grounded in handcrafted rules or simpler statistical methods. The short answer lies in their superior ability to model and leverage context. Older rule-based approaches were often rigid, relying on explicitly encoded grammars or dictionaries that could not easily adapt to the countless linguistic nuances found in real-world discourse. Likewise, early statistical models—such as basic n-gram approaches—were constrained by fixed windows of context and struggled to capture long-range dependencies. In contrast, modern LLMs incorporate advanced architectures like the transformer (Vaswani et al., 2017), which employ attention mechanisms to dynamically weigh and integrate information across entire sequences of text (see Chapter 1 and the technical annex at the end of the book). This richer contextual awareness allows them not only to generate more natural-sounding language but also to handle semantic ambiguities and rare usages with a level of finesse previously out of reach.

To illustrate the importance of contextual modeling, it is helpful to revisit one of the earliest and most influential examples in the field: Bar-Hillel's analysis of the word 'pen.' This classic case highlights how context is essential for resolving ambiguity. In a sentence like 'Little John was looking for his toy box... The box was in the pen,' the intended meaning of pen—a small enclosed area for a child— is uncommon and cannot be inferred from frequency alone. Bar-Hillel, writing in the late 1950s, emphasizes how early machine translation systems struggled with such cases, as they lacked the contextual understanding needed to disambiguate rare or subtle senses. His example continues to serve as a touchstone for the challenges of lexical ambiguity in computational models.

I now claim that no existing or imaginable program will enable an electronic computer to determine that the word pen in the given sentence within the given context has the second of the above meanings, whereas every reader with a sufficient knowledge of English will do this 'automatically.' Incidentally, we realize that the issue is not one that concerns translation proper, i.e., the transition from one language to another, but a preliminary stage of this process, or, the determination of the specific meaning in context of a word which, in isolation, is semantically ambiguous (relative to a given target-language, if one wants to guard oneself against the conceivable though extremely unlikely case that the target-language contains a word denoting both the same writing utensil and an enclosure where children can play). (Bar-Hillel, 1959)

In the 1960s, the task in question was deemed impossible to solve; as Bar-Hillel observed, 'no existing or imaginable program' at the time could address it. Consequently, machine translation funding dwindled and nearly dried up for decades following Bar-Hillel's report. The ALPAC report (1965; see also Hutchins, 2003) authored by a panel of experts, echoed Bar-Hillel's conclusions, asserting not only the infeasibility of the task but also highlighting the lack of a viable market for machine translation. These observations had a significant impact, effectively stalling progress in the field for years.

Amid this funding crisis and uncertainty, debates over the role of semantics in language processing—and linguistics more broadly—gained new prominence. Noam Chomsky's transformational-generative grammar (1957, 1965), dominant at the time, largely dismissed semantics, treating the lexicon as a repository of irregularities rather than a structured system worthy of deeper analysis. Chomsky's approach prioritized syntactic structures and universal rules, implying that meaning played little role in the underlying generative mechanisms. This stance provoked opposition from scholars who argued that semantics was fundamental to understanding both language and translation. The resulting divide between these positions sparked what would later be dubbed the 'linguistic wars,' a conflict meticulously chronicled by R. A. Harris (1993). As support for practical applications like machine translation waned, theoretical battles over syntax and semantics raged, setting the stage for decades of contention and shaping the field's trajectory.

Among Chomsky's notable opponents was George Lakoff, who became a central figure in the generative semantics movement.

Lakoff (1971) and his colleagues sought to integrate semantic considerations directly into grammatical rules, arguing that meaning could not be divorced from the structure of sentences. This approach represented a dramatic departure from Chomsky's rigid syntactic framework and proposed that the underlying structures of language were inherently semantic in nature. James McCawley, another prominent generative semanticist, further developed these ideas by offering alternative analyses of grammatical phenomena that relied on semantic interpretations rather than purely syntactic derivations (McCawley, 1971). Meanwhile, John R. Ross, initially a Chomsky ally, broke away to explore a more meaning-driven approach to grammar, developing influential concepts like 'performative analysis' that highlighted the importance of pragmatic and semantic information (Ross, 1970). These scholars collectively challenged the Chomskyan paradigm, paving the way for alternative models that sought a deeper integration of syntax, semantics, and pragmatics in understanding human language.

However, even for these approaches, a proper formalization of word meaning remained elusive. A central question was whether, and how, sense inventories such as those found in traditional dictionaries could be applied effectively. The challenge of assigning the correct sense to a given word in context—known as word sense disambiguation (WSD)—emerged as one of the major tasks in NLP during the 1990s and 2000s (Navigli, 2009; Yarowsky, 2001). Efforts like SenseEval, later known as SemEval,[2] provided an annual venue to evaluate systems tackling WSD, measuring their progress against manually annotated gold standards. Ironically, one of the driving figures behind this evaluation effort, Adam Kilgarriff, had authored a notable paper titled 'I Don't Believe in Word Senses,' which questioned the traditional, rigid notion of discrete word meanings (Kilgarriff, 1997).

These shared evaluation efforts helped formalize the problem of WSD but they also revealed the constraints of prevailing methods.

---

[1] https://web.eecs.umich.edu/~mihalcea/senseval.

[2] https://semeval.github.io.

The dominant approach during those years relied on predefined inventories of word senses, often derived from resources such as WordNet[3] (Fellbaum, 1998). However, the results—both in terms of system performance and interannotator agreement—remained underwhelming. This led researchers to question whether these inventories genuinely captured the nuances of meaning encountered in natural language. For instance, Erk et al. (2013) proposed a graded approach to word senses, suggesting that word meanings are not neatly separated but instead exist on a continuum. In their view, word sense boundaries are fluid, overlapping, and influenced by context in complex ways. This perspective pointed toward the need for more nuanced and flexible models of meaning—models that could better reflect the dynamic, context-dependent nature of language use.

But, in the end, what is context? The fundamental problem for computational linguists has been that 'context' is not a simple, formal concept—no single theory has fully encapsulated its complexity. Since at least the 1950s, a multitude of frameworks have attempted to formalize context, drawing on fields ranging from linguistics (Harris, 1954; Van Dijk and Kintsch, 1983) to cognitive science (Minsky, 1975) and philosophy of language (Stalnaker, 1999). Yet, none of these approaches yielded a fully satisfactory, operational model of context for computational systems. For a computational language model, there is in fact no formal definition of context beyond the list of preceding tokens in a text stream; the model simply conditions its next output on this sequence. These models manage to process such a notion of context with remarkable efficiency by leveraging attention-based mechanisms (Vaswani et al., 2017) capable of tracking dependencies across thousands of tokens. As a result, memory constraints are no longer a major obstacle, since modern architectures can take into account extensive portions of prior discourse during analysis.

---

[3] WordNet: A lexical database for English (Version 3.0). Cognitive Science Laboratory, Princeton University. https://wordnet.princeton.edu.

In addition to their adept handling of context, LLMs benefit from several other advances over previous systems. First, they are trained on vastly larger and more diverse datasets, encompassing web texts, books, and specialized domains, which helps them capture rare linguistic patterns and domain-specific knowledge. Second, they utilize deep neural architectures capable of learning complex representations at multiple layers, far outstripping the capacity of older statistical or rule-based models. Third, the rise of massive parallel computing (particularly with GPUs and TPUs) enables these models to train efficiently on billions of parameters, further enhancing their generalization capabilities (Brown et al., 2020; Kaplan et al., 2020). By combining a deep contextual understanding with immense training corpora and powerful computational infrastructures, LLMs outperform earlier approaches in fluency, adaptability, and semantic accuracy, thereby validating Bar-Hillel's early intuition that context—and the ability to handle it—lies at the heart of meaningful language processing.

2 Reassessing linguistic theory in light of large language models

The rise of LLMs has profoundly affected not only the field of natural language processing but also the fundamental assumptions of linguistic theory. Traditionally, linguistics—especially within the generative tradition—has emphasized formal, symbolic rules as the backbone of language competence. Yet the remarkable empirical success of LLMs suggests that much of linguistic knowledge might instead be modeled through powerful statistical learning mechanisms grounded in exposure to language use. This shift raises deep questions about the adequacy of rule-centric theories and invites a broader reassessment of linguistic explanation.

Three recent contributions frame this debate in particularly illuminating ways.

First, Marco Baroni (2022) critically examines how LLMs might be relevant to linguistic theorizing. Baroni cautions against treating deep neural models as transparent cognitive mirrors, yet argues that their ability to capture intricate linguistic patterns without explicit symbolic rules challenges long-standing assumptions about the architecture of linguistic competence. He proposes that, rather than dismissing LLMs as mere engineering artifacts, linguists should see them as algorithmic linguistic theories capable of making explicit, testable predictions about utterance acceptability. However, Baroni identifies major obstacles to their integration into mainstream linguistics, such as a 'low commitment to models'—where research often prioritizes practical NLP performance over theoretical insight—and a 'lack of mechanistic understanding' of how these models actually generalize. In his view, LLMs could inspire a more empirical and predictive linguistics, provided they are analyzed with the same rigor traditionally applied to symbolic frameworks.

Second, Steven Piantadosi (2024) offers a direct and forceful challenge to the Chomskyan tradition. He argues that LLMs fundamentally undermine the necessity of positing innate, domain-specific structures like universal grammar (UG). For Piantadosi, the ability of these models to acquire complex linguistic structures solely from large-scale textual input—without prespecified grammatical rules or biological priors—demonstrates that language learning is achievable through general-purpose mechanisms. This, he contends, refutes the poverty of the stimulus argument, one of the central pillars of generative linguistics. Piantadosi further highlights how LLMs integrate syntax and semantics in gradient, continuous vector spaces, employ probability and information-theoretic principles, and develop hierarchical representations organically through training—all features that contradict core Chomskyan tenets. While he acknowledges that the generative tradition contributed valuable concepts like island constraints, he insists that LLMs show these phenomena can emerge from experience-driven learning rather than innate specification.

Third, Roni Katzir (2023) offers a more cautious perspective, warning against framing the debate as a simple opposition between symbolic formalisms and data-driven models. Katzir argues that, while LLMs reveal important empirical patterns, they fail to provide a convincing account of why human language takes its specific forms or how linguistic competence differs from performance. He points out that LLMs often misjudge grammatical acceptability, preferring statistically probable but theoretically ungrammatical continuations. Moreover, he questions whether LLMs genuinely capture deep linguistic universals, highlighting cases such as the conservativity of quantifiers or the absence of palindromic phonological sequences, which he argues cannot be fully explained by pattern-matching alone. Katzir stresses that formal tools and theoretical rigor remain indispensable for understanding the underlying architecture of linguistic competence, even as LLMs enrich our empirical palette.

Taken together, these three perspectives illuminate a field in transition. On the one hand, LLMs challenge entrenched assumptions about the necessity of explicit, innate rule systems, showing that rich exposure to language can generate complex linguistic behavior. On the other hand, they leave unresolved questions about the cognitive and formal basis of grammar, reminding us that performance is not competence, and that empirical successes do not automatically yield explanatory depth (Millière, 2025).

This moment of convergence invites linguists to rethink the boundaries of their frameworks. Instead of treating LLMs as purely external engineering tools, they might be considered experimental probes that expose latent assumptions within linguistic theory. Their success suggests that phenomena once considered peripheral—such as frequency effects, lexical richness, and context-sensitivity—may in fact be central to our understanding of language competence. Engaging rigorously with the capacities and limitations of LLMs does not reject linguistic theory, but rather offers an opportunity to expand, refine, and empirically test it in light of new modeling paradigms.

3 And in practice? Lessons for linguistic theory from large language models

As we have just seen, while LLMs do not replicate human cognitive processes, their impressive ability to produce coherent language from statistical patterns challenges traditional assumptions about symbolic or innate frameworks and invites a reassessment of how linguistic knowledge is organized. In this section, we examine how LLMs illuminate core aspects of linguistic form and usage, not by modeling the human mind but by revealing patterns latent in language as a social and statistical phenomenon.

3.1 Beyond context, the role of frequency in language

As discussed in the previous section, context plays a central role in both language understanding and production. Closely intertwined with context, yet warranting distinct emphasis, is the role of frequency effects in shaping linguistic behavior. Frequency patterns influence not only the occurrence of individual words but also extend across various levels of linguistic organization, including syntactic structures, word senses, and collocations. A well-documented manifestation of this is the generalized Zipfian distribution (Zipf, 1949), which captures how a small subset of words appears with extremely high frequency, while the vast majority are used rarely. Importantly, this distributional tendency is not confined to vocabulary but applies more broadly to language patterns at multiple levels.

For instance, most words tend to have a primary sense that overwhelmingly dominates in frequency over secondary or peripheral meanings. This observation dovetails with our earlier discussion on the fluidity of meaning, and how sharp demarcations between distinct word senses may be more of a theoretical convenience than an accurate reflection of language use. Similarly, certain syntactic constructions—such as subject–verb agreement or canonical word order—are highly regular and frequent, whereas others, especially more marked or complex structures, exhibit lower frequency and greater variability. These frequency-driven tendencies are not merely descriptive artifacts; rather, they exert a formative influence on both human language acquisition and the learning processes of LLMs, guiding how patterns are internalized, generalized, and prioritized.

A significant body of linguistic research has underscored the need to incorporate statistical and frequency-based information into theoretical accounts of language. Traditional formalist frameworks, particularly those associated with generative grammar (Chomsky, 1957, 1965, 1995), have largely foregrounded the role of innate principles, categorical rules, and binary grammaticality judgments. While these models have offered profound insights into the architecture of linguistic competence, they tend to treat frequency effects as epiphenomenal or irrelevant to the core of linguistic knowledge.

In contrast, alternative approaches such as construction grammar (Goldberg, 1995; 2006), usage-based linguistics (Bybee, 2010; Tomasello, 2003), and emergentist models (MacWhinney, 1999) argue that linguistic structures arise from patterns of use, heavily shaped by experience and frequency. Studies in corpus linguistics (Biber et al., 1998) reinforce this view, demonstrating that language exhibits gradient, probabilistic patterns rather than strictly categorical distinctions. Likewise, findings in psycholinguistics indicate that frequency and predictability significantly influence cognitive processes such as parsing, lexical access, and production fluency (Ellis, 2002; Hale, 2001; Jurafsky, 1996; Levy, 2008).

LLMs, by design, align more closely with these frequency-based approaches. Their learning mechanism involves the statistical assimilation of vast amounts of textual data, allowing them to capture fine-grained regularities, rare patterns, and distributional nuances. This empirical success invites a reevaluation of linguistic theories: Rather than viewing frequency effects as peripheral, we may need to recognize them as central to both the acquisition and use of language.

3.2 The centrality of rich and continuous lexical representations (beyond abstract formal rules)

One of the most fundamental aspects of language—often underestimated in formal linguistic theories—is the richness of lexical representations and the wealth of information attached to them. In the case of LLMs, linguistic knowledge is not encoded primarily through a small set of universal syntactic rules but rather through dense, high-dimensional representations of words and tokens, as we saw in Chapter 1. These representations capture not only semantic content but also syntactic behavior, collocational tendencies, and pragmatic associations. Crucially, these representations are continuous rather than symbolic, allowing for fine-grained gradations of meaning and usage, rather than rigid categorical distinctions.

This emphasis on lexical specificity stands in contrast to many contemporary formal linguistic frameworks, particularly those influenced by generative grammar, which have historically aimed to distill language into minimal, abstract rule systems operating independently of the lexicon (Chomsky, 1995). However, LLMs reveal that much of what constitutes linguistic competence is inherently word-specific and usage-dependent, challenging the assumption that general principles alone can account for the richness of natural language.

Even constructions often considered as prototypical examples of abstract syntactic rules—such as the formation of passive voice—are deeply intertwined with lexical properties. For instance, a verb must first be recognized as transitive before passivization is possible (and even so, there are nuances. For example, while 'The examiner failed the candidate' ⟹ 'The candidate was failed by the examiner' is perfectly acceptable, 'He failed his team when they needed him most'. ⟹ *'His team was failed by him'. is marked, sounds strange, and is usually avoided in normal English usage. The latter has an overly formal or even legalistic tone) (Huddleston and Pullum, 2002). Furthermore, in other languages such as French, the specific realization of passive constructions can vary depending on both the verb and the context. While the agent in a passive sentence is conventionally introduced by 'par' in French (equivalent to 'by' in English, e.g., 'Ce livre a été écrit par Proust' 'This book was written by Proust'), alternative prepositional phrases are frequently employed with verbs of 'liking' (*aimer, adorer, apprécier,* as seen in 'Albertine a été aimée de Proust' 'Albertine was loved by Proust') (Straub, 1974). Such examples highlight how even 'general' rules are language-dependent and conditioned by lexical and contextual factors—suggesting that linguistic regularities are often emergent properties of rich lexical information rather than reflections of universally applicable rules.

This tension between abstract syntactic principles and lexical knowledge has long shaped debates within linguistics (Harris, 1993). Traditional rule-based approaches have tended to marginalize lexical richness, treating the lexicon as a relatively arbitrary list of items over which universal principles operate. Yet both empirical studies of language use and the performance of LLMs point to a more lexically driven view of linguistic competence, where much of language's generative power arises not from formal rules but from densely interconnected, usage-based lexical patterns.

In computational linguistics, efforts to systematically encode and structure lexical knowledge have a rich history. Early projects such as WordNet (Fellbaum, 1998) and FrameNet (Fillmore, 2003) sought to formalize word meanings and their relationships in structured, machine-readable formats. WordNet organized words into hierarchically arranged synsets, encoding semantic relations like synonymy, hyponymy, and antonymy. FrameNet took a broader approach, associating lexical items with conceptual frames—schematic representations of typical situations and participant roles.

While pioneering, these projects faced inherent limitations. First, they relied on manually defining word senses and relations, an approach that struggled to keep pace with the fluidity, variability, and context-dependency of actual language use. Second, lexical relations are rarely as neat and hierarchical as early models assumed. The meaning of a word frequently shifts depending on context, cooccurrence patterns, and cultural background—dimensions difficult to capture through fixed, symbolic structures.

In this respect, the move to continuous, distributed lexical representations in LLMs marks a significant epistemological shift. Rather than treating lexical knowledge as a finite, manually curated set of discrete entries, LLMs learn to represent words as points in a high-dimensional vector space, shaped by patterns of cooccurrence across vast corpora. This approach captures subtle gradations of meaning and syntactic behavior, allowing models to generalize flexibly across contexts without relying on predefined rules or exhaustive manual effort.

Philosophically, this shift invites reconsideration of long-standing assumptions about the nature of linguistic knowledge. Formalist theories often valorize systematicity, abstraction, and rule-governed competence, drawing parallels to formal systems in logic or mathematics. However, LLMs demonstrate that linguistic competence may be better understood as emergent, probabilistic, and deeply embedded in lexically specific knowledge—a view that resonates with empiricist and holistic traditions in the philosophy of language and cognitive science. For instance, Quine (1960) challenged the idea that some statements are true solely by virtue of their meaning, independent of empirical facts. He argued instead that our understanding of meaning depends on how language is used in practice, and that there can be no single, definitive way to translate between languages (what he called 'the indeterminacy of translation').

Furthermore, the reliance on continuous representations challenges the symbolic paradigm traditionally favored in cognitive models of language. It raises questions about whether discrete symbolic manipulation is truly necessary to account for human linguistic ability, or whether cognitive systems may similarly rely on high-dimensional, subsymbolic representations shaped by experience and interaction with linguistic input.

4 Large language models and language acquisition

One of the most striking findings in recent research on LLMs is that the ability to generate grammatically well-formed sentences—and even coherent, structured texts—can emerge purely from exposure to large-scale textual data, without any explicit grammatical rules being preprogrammed. This suggests that much of what we consider linguistic competence, including syntax, semantics, and discourse-level organization, can be acquired through statistical learning from input alone. LLMs achieve this by capturing distributional patterns across vast corpora—hundreds of billions of words drawn from books, articles, websites, and other written sources. Rather than applying predefined grammatical principles, these models learn by adjusting internal representations based on cooccurrence statistics and structural regularities. This success lends empirical weight to usage-based and emergentist theories of language learning (Bybee, 2010; Tomasello, 2003), which emphasize that linguistic competence arises from experience, frequency, and associative learning, rather than innate syntactic templates.

However, the learning paradigm underlying LLMs diverges sharply from what is currently understood about human language acquisition. Human infants are exposed to only a minute fraction of the input LLMs receive, yet they appear to achieve mastery of complex linguistic structures within just a few years. This discrepancy raises important questions about what language learning involves, and whether some form of innate predisposition is necessary. For decades, influential hypotheses rooted in generative grammar—most notably Chomsky's proposal of UG—have suggested that children possess domain-specific, biologically prespecified knowledge to compensate for the so-called poverty of the stimulus (Chomsky, 1980). According to this view, the linguistic input available to children may be too sparse and noisy to fully account for the richness of the grammar they acquire—though this position remains contested within the field.

While the view that innate mechanisms are required remains debated, the empirical performance of LLMs further complicates the picture. These models demonstrate that many aspects of linguistic competence—including syntactic structure, agreement, and coherence—can emerge from exposure to unannotated input, without innate constraints. At the same time, their differences from human learners are equally instructive. LLMs are trained nonincrementally, consuming enormous datasets in bulk, without developmental stages, interactive feedback, or social engagement. Human learners, by contrast, acquire language gradually, starting with simple communicative acts and building toward complex grammatical constructions. This process is scaffolded by joint attention, embodied interaction, and multimodal sensory input—factors that are entirely absent from LLM training paradigms. As Saxton (2017) emphasizes, language acquisition in children unfolds within a rich sociocognitive environment, shaped by feedback, negotiation of meaning, and real-time interaction.

These differences raise important theoretical questions. If language learning does not rely on innate syntactic structures, what innate capacities might underlie it? An alternative view, proposed by Christiansen and Chater (2016), suggests that language acquisition is supported not by a specialized language faculty but by domain-general cognitive abilities—such as pattern recognition, working memory, attention, and social reasoning—which allow learners to infer structure from limited input and context. From this perspective, linguistic knowledge is not hardwired but emerges from the interaction between cognitive processes and structured experience.

Philosophically, this shift invites a reconsideration of nativist versus empiricist approaches to language. LLMs lend empirical support to the idea that rich linguistic behavior can emerge from statistical generalization over input, aligning with broader empiricist traditions in epistemology. For example, Churchland (1995) proposes a connectionist model of cognition in which knowledge is encoded in distributed neural representations rather than symbolic rules. While LLMs do not replicate the full cognitive and social context of human learning, their success challenges the assumption that formal rule systems are a necessary precondition for linguistic knowledge. At the same time, their limitations—particularly their dependence on massive data and lack of embodied, interactive learning—highlight the unique richness of human language acquisition, which likely depends on a confluence of statistical, cognitive, and social factors.

5 Limitations

To conclude this chapter, it is essential to highlight two key points regarding the evolving relationship between linguistic theory and LLMs, as well as the challenges that remain.

First, scientific progress is rarely linear, and every theoretical and methodological framework we have examined—whether grounded in formal rules, statistical patterns, or neural networks—has played a critical role in advancing our understanding of language. One must recognize the immense and lasting influence of Chomskyan linguistics, even as some assumptions within this tradition are increasingly questioned in light of modern LLMs.

Chomsky's contributions fundamentally shaped the study of language, not only by introducing the concept of UG but also by establishing formal rigor in linguistic theorizing. His insistence on the competence/performance distinction and the notion of language as an internally coherent, rule-governed system provided a framework that moved linguistics toward a systematic, scientific discipline. Even today, aspects of generative grammar continue to offer valuable insights. For instance, the hierarchical structure of syntax, such as X-bar theory, has highlighted the recursive and compositional nature of human language—properties that surface in LLM outputs but are not explicitly encoded in their architecture. Understanding the deep structural properties of sentences, independent of surface word order, remains a theoretical lens through which certain linguistic phenomena can be more clearly interpreted, even when analyzing LLM outputs.

Thus, while LLMs demonstrate that large-scale statistical learning can approximate many linguistic competencies, they do not render Chomskyan theory obsolete. Instead, they invite a reassessment of which aspects of linguistic competence are best modeled by innate principles, and which emerge from exposure to language use. As Baroni (2022) and others have argued, LLMs should be seen not as definitive answers but as new tools that can inform, refine, and challenge traditional linguistic models.

Second, and perhaps more pressing, is the fact that, despite their empirical success, LLMs remain opaque systems whose inner workings are only partially understood. The ability of LLMs to capture linguistic regularities is undeniable, but how exactly linguistic properties are encoded within these models remains largely an open question. In contrast to symbolic linguistic theories, where explicit rules and categories can be inspected and debated, the knowledge encoded in LLMs is distributed across millions or even billions of parameters, arranged in ways that resist straightforward interpretation.

Researchers have made significant efforts to shed light on these mechanisms. Techniques such as probing classifiers (Conneau et al., 2018), ablation studies, and emerging approaches in mechanistic interpretability (Nanda et al., 2023) have been deployed to investigate how syntactic, semantic, and even pragmatic information is represented internally. However, these efforts have revealed both the promise and the limits of current understanding of LLMs. While it is possible to find correlations between certain model components and linguistic properties (e.g., subject–verb agreement, constituent structure), we are still far from a comprehensive account of what linguistic knowledge is encoded, how it is distributed, and the extent to which it mirrors human linguistic competence.

For example, although LLMs often perform well on standard syntactic benchmarks, they still exhibit errors in handling negation, coreference resolution, and logical entailment—phenomena that require deeper structural and inferential reasoning. In particular, negation frequently poses challenges: Models sometimes ignore it altogether or misinterpret its scope, producing outputs that contradict the intended meaning, although this is becoming less true with recent advancements—larger models tend to be more robust than smaller ones. The importance of scale remains both intriguing and revealing. It suggests that certain capabilities emerge and are more finely tuned as the model is exposed to larger amounts of data. At the same time, this highlights a fundamental limitation of such systems: While they may gradually converge toward more optimal solutions, they do so without relying on well-grounded a priori theories, and they may lack robust internal mechanisms to represent certain logical and compositional properties of language.

As Katzir (2023) emphasizes, this opacity is not a trivial matter. Without clearer interpretability, it becomes difficult to diagnose model failures, trace their decision-making processes, or reliably assess the linguistic competence they exhibit. Katzir argues for the continued importance of formal models and theoretical clarity, cautioning against the temptation to view LLM performance as self-explanatory.

Consequently, a better theoretical and empirical understanding of linguistic phenomena within LLMs remains necessary. Moving forward, research must focus not only on improving LLM performance but also on developing principled methodologies to interpret and explain their behavior. This will likely require interdisciplinary collaboration, integrating insights from theoretical linguistics, cognitive science, and machine learning. It will involve refining probing methods, enhancing model transparency, and perhaps even developing new architectures that blend the strengths of formal linguistic theory with data-driven learning.

Much like earlier linguistic and computational models, LLMs should be subjected to rigorous critical analysis, with their limitations as carefully examined as their successes. Understanding how these models represent language—and how they fail to do so—will be crucial for building systems that are not only effective but also theoretically insightful and cognitively plausible.

Ultimately, the story of linguistic theory and LLMs is one of cumulative progress. Past frameworks, from Chomskyan formalism to corpus-based approaches, have each contributed essential pieces to our current knowledge. LLMs represent the latest chapter in this ongoing narrative—not as a final answer but as a powerful, complex tool that both challenges and enriches our understanding of language.

References

ALPAC. (1966). *Languages and machines: Computers in translation and linguistics. A report by the Automatic Language Processing Advisory Committee, Division of Behavioral Sciences, National Academy of Sciences, National Research Council.* National Academy of Sciences,

National Research Council. https://nap.nationalacademies.org /resource/alpac_lm/ARC000005.pdf

Bar-Hillel, Y. (1959). *Report on the state of machine translation in the United States. Technical report no. 1. Prepared for the U.S. Office of Naval Research, Information Systems Branch, Jerusalem, Israel, 1959.* https://aclanthology.org/www.mt-archive.info/Bar-Hillel-1959.pdf

Baroni, M. (2022). On the proper role of linguistically-oriented deep net analysis in linguistic theorizing. In S. Lappin (Ed.), *Algebraic systems and the representation of linguistic knowledge* (pp. 5–22). Taylor & Francis.

Biber, D., Conrad, S., and Reppen, R. (1998). *Corpus linguistics: Investigating language structure and use.* Cambridge University Press.

Brown, T., Mann, B., Ryder, N., Subbiah, M., Kaplan, J., Dhariwal, P., Neelakantan, A., Shyam, P., Sastry, G., Askell, A., Agarwal, S., Herbert-Voss, A., Krueger, G., Henighan, T., Child, R., Ramesh, A., Ziegler, D., Wu, J., Winter, C. … Amodei, D. (2020). Language models are few-shot learners. *Advances in Neural Information Processing Systems (NeurIPS), 33,* 1877–1901. https://dl.acm.org/doi/abs /10.5555/3495724.3495883

Bybee, J. (2010). *Language, usage and cognition.* Cambridge University Press.

Chomsky, N. (1957). *Syntactic structures.* Mouton de Gruyter.

Chomsky, N. (1965). *Aspects of the theory of syntax.* The MIT Press.

Chomsky, N. (1980). *Rules and representations.* Columbia University Press.

Chomsky, N. (1995). *The minimalist program.* The MIT Press.

Christiansen, M. H., and Chater, N. (2016). *Creating language: Integrating evolution, acquisition, and processing.* The MIT Press.

Churchland, P. M. (1995). *The engine of reason, the seat of the soul: A philosophical journey into the brain.* The MIT Press.

Conneau, A., Kruszewski, G., Lample, G., Barrault, L., and Baroni, M. (2018). What you can cram into a single $&!#* vector: Probing sentence embeddings for linguistic properties. In *Proceedings of the 56th Annual Meeting of the Association for Computational Linguistics (Volume 1: Long Papers), Melbourne, Australia* (pp. 2126–2136). Association for Computational Linguistics. https://aclanthology.org /P18-1198/

Ellis, N. C. (2002). Frequency effects in language processing: A review with implications for theories of implicit and explicit language acquisition. *Studies in Second Language Acquisition, 24*(2), 143–188.

Erk, K., McCarthy, D., and Gaylord, N. (2013). Measuring word meaning in context. *Computational Linguistics, 39*(3), 511–554. https://doi.org/10.1162/COLI_a_00142

Fellbaum, C. (1998). *WordNet: An electronic lexical database.* The MIT Press. https://doi.org/10.7551/mitpress/7287.001.0001

Fillmore, C. J. (2003). *Form and meaning in language. Volume II: Papers on semantic roles.* CSLI.

Goldberg, A. E. (1995). *Constructions: A construction grammar approach to argument structure.* University of Chicago Press.

Goldberg, A. E. (2006). *Constructions at work: The nature of generalization in language.* Oxford University Press.

Hale, J. (2001). A probabilistic Earley parser as a psycholinguistic model. In *Second Meeting of the North American Chapter of the Association for Computational Linguistics (NAACL), Carnegie Mellon University, Pittsburgh, USA.* Association for Computational Linguistics. https://aclanthology.org/N01-1021.pdf

Harris, Z. (1954). Distributional structure. *Word, 10*(2–3), 146–162.

Harris, R. A. (1993). *The linguistics wars: Chomsky, Lakoff and the battle over deep structure.* Oxford University Press.

Huddleston, R., and Pullum, G. K. (2002). *The Cambridge grammar of the English language.* Cambridge University Press.

Hutchins, J. (2003). ALPAC: The (in)famous report. In S. Nirenburg, H. L. Somers and Y. Wilks (Eds.), *Readings in machine translation.* The MIT Press. https://doi.org/10.7551/mitpress/5779.001.0001

Jurafsky, D. (1996). A probabilistic model of lexical and syntactic access and disambiguation. *Cognitive Science, 20*(2), 137–194.

Kaplan, J., McCandlish, S., Henighan, T., Brown, T. B., Chess, B., Child, R., Gray, S., Radford, A., Wu, J., and Amodei, D. (2020). Scaling laws for neural language models. arXiv preprint. https://arxiv.org/abs/2001.08361

Katzir, R. (2023). Why large language models are poor theories of human linguistic cognition: A reply to Piantadosi. *Biolinguistics, 17,* e13153.

Kilgarriff, A. (1997). I don't believe in word senses. *Computers and the Humanities, 31,* 91–113. https://doi.org/10.1023/A:1000583911091

Lakoff, G. (1971). On generative semantics. In D. Steinberg and L. Jakobovits (Eds.), *Semantics: An interdisciplinary reader in philosophy, linguistics and psychology* (pp. 232–296). Cambridge University Press.

Levy, R. (2008). Expectation-based syntactic comprehension. *Cognition, 106*(3), 1126–1177.

McCawley, J. D. (1971). Where do noun phrases come from? In D. Steinberg and L. Jakobovits (Eds.), *Semantics: An interdisciplinary reader in philosophy, linguistics and psychology* (pp. 217–231). Cambridge University Press.

MacWhinney, B. (1999). The emergence of language from embodiment. In B. MacWhinney (Ed.), *The emergence of language* (pp. 213–256). Lawrence Erlbaum.

Millière, R. (2025). Language models as models of language. In *The Oxford handbook of the philosophy of linguistics* (Preprint available on the author's website). Oxford University Press. https://raphaelmilliere .com/pdfs/milliereLanguageModelsModelsforthcoming.pdf

Minsky, M. (1975). A framework for representing knowledge. In P. H. Winston (Ed.), *The psychology of computer vision* (pp. 211–277). McGraw-Hill.

Nanda, N., Chan, L., Lieberum, T., Smith, J., and Steinhardt, J. (2023). Progress measures for grokking via mechanistic interpretability. In *The Eleventh International Conference on Learning Representations (ICLR), Kigali, Rwanda.* https://iclr.cc/virtual/2023/oral/12572

Navigli, R. (2009). Word sense disambiguation: A survey. *ACM Computing Surveys, 41*(2), 1–69. https://doi.org/10.1145/1459352.1459355

Piantadosi, S. T. (2024). Modern language models refute Chomsky's approach to language. In E. Gibson and M. Poliak (Eds.), *From fieldwork to linguistic theory: A tribute to Dan Everett* (pp. 353–414). Language Science Press.

Quine, W. V. O. (1960). *Word and object.* The MIT Press.

Ross, J. R. (1970). On declarative sentences. In R. A. Jacobs and P. S. Rosenbaum (Eds.), *Readings in English transformational grammar* (pp. 222–272). Ginn and Company.

Saxton, M. (2017). *Child language: Acquisition and development* (2nd ed.). Sage Publications.

Stalnaker, R. (1999). *Context and content: Essays on intentionality in speech and thought.* Oxford University Press.

Straub, S. A. (1974). The passive prepositions de and par. *The French Review, 47*(3), 583–593.

Tomasello, M. (2003). *Constructing a language: A usage-based theory of language acquisition.* Harvard University Press.

Van Dijk, T. A., and Kintsch, W. (1983). *Strategies of discourse comprehension.* New York: Academic Press.

Vaswani, A., Shazeer, N., Parmar, N., Uszkoreit, J., Jones, L., Gomez, A. N., Kaiser, Ł., and Polosukhin, I. (2017). Attention is all you need. *Advances in Neural Information Processing Systems (NeurIPS), Long Beach (pages 5998–6008), 30.* https://arxiv.org/abs/1706.03762

Yarowsky, D. (2001). Word sense disambiguation. In R. Dale, H. Moisl, and H. Somers (Eds.), *Handbook of natural language processing* (pp. 629–654). Marcel Dekker.

Zipf, G. K. (1949). *Human behavior and the principle of least effort: An introduction to human ecology.* Addison-Wesley.

CHAPTER 3

Large language models and the future of writing

In earlier sections (esp. in Chapter 1), we explored how linguistic and literary theories—more specifically structuralism and poststructuralism—challenged traditional notions of authorship, originality, and meaning. Thinkers such as Barthes and Foucault emphasize that texts are not expressions of a singular, sovereign author but rather assemblages of prior discourses, formed through citation, repetition, and cultural convention (Underwood, 2023). In this view, the figure of the 'author' becomes less a source of meaning than a function of discourse itself—a placeholder within a broader system of language (as already defined by Saussure in his distinction between language, *langue*, and speech, *parole*). Language precedes and exceeds the individual writer, shaping thought and constraining expression.

Large language models (LLMs) such as ChatGPT instantiate this theoretical insight with remarkable fidelity. Trained on massive corpora of human-generated text, these systems produce coherent and plausible continuations of discourse by statistically reproducing patterns they have encountered during training (see

How to cite this book chapter:
Poibeau, T. 2025. *Understanding Conversational AI: Philosophy, Ethics and Social Impact of Large Language Models.* Pp. 65–82. London: Ubiquity Press. DOI: https://doi.org/10.5334/bde.d. License: CC BY-NC 4.0

Chapters 1 and 2). In a literal sense, they generate texts without authors—outputs that are fluent, plausible, and often informative, yet lack a clear point of origin or intentionality. What was once a theoretical idea—that writing is an act of recombination rather than creation—has become a common reality.

This chapter turns to the practical implications of this shift. How are LLMs already transforming writing practices across different domains—from encyclopedic knowledge production on Wikipedia to scientific communication and academic publishing, to the everyday genres of emails, resumes, and blog posts? How should we detect, regulate, or attribute machine-generated text, especially when it is edited or modified by human users? And what are the consequences of these transformations for education, authorship, and the future of written expression?

By focusing on these questions, the chapter aims to move beyond abstraction and address the material changes underway in writing culture. In doing so, it confronts a deeper uncertainty: What does it mean to write in a world where machines can do so as well?

1 Authorship revisited: large language models and the fragmentation of the writing subject

If structuralist and poststructuralist thinkers destabilized the authority of the individual author, LLMs have now brought that destabilization into the domain of practical writing. No longer confined to theoretical discourse, the idea that a text may lack a coherent or intentional author has become a concrete feature of contemporary textual production. Systems like ChatGPT can now generate fluent, contextually appropriate prose across a wide range of genres, without possessing experience, consciousness, or even a stable perspective (we will discuss issues like consciousness in more detail in the following chapters).

Foucault's (1977 [1969]) rethinking of authorship as a *function*—a regulatory principle rather than a personal identity—finds new relevance here. In *What Is an Author?* Foucault argues that authorship is not a natural or universal category but a juridical and discursive function that historically emerged to regulate the circulation and reception of certain texts—particularly those deemed valuable, dangerous, or punishable. In the age of LLMs, this author function becomes deeply unstable. When a user generates a legal memo, a scientific abstract, or a narrative paragraph via a prompt, the notion of the author becomes blurred and subject to challenge—distributed between the user, the model, and its designers.

Legal frameworks offer no straightforward answer. Copyright law in most jurisdictions remains rooted in a conception of authorship that presumes human creativity and intentionality. The US Copyright Office, for instance, has explicitly stated that works generated solely by AI systems are ineligible for protection, reaffirming that authorship must originate in a natural person (US Copyright Office, 2023). At the same time, legal disputes are emerging around the use of copyrighted training data—raising questions about whether outputs reflect 'transformative use' or unauthorized reproduction. If a model reproduces the style, syntax, or argumentation of a given author, does it amount to pastiche, plagiarism, or legal infringement?

Philosophically, this tension echoes earlier debates about the ontology of the work of art. Walter Benjamin's (1969 [1935]) essay *The Work of Art in the Age of Mechanical Reproduction* argues that mass technologies devalued the 'aura' of the original artwork, eroding its uniqueness, context, and authority. LLMs extend this logic: They automate not only the reproduction of form but the generation of new text, undermining the notion of originality as such. The digital text, like the mechanically reproduced image, is detached from its origin, infinitely iterable, and stripped of any intrinsic singularity. Yet, in contrast to Benjamin's case, what is reproduced here is not a prior work but a statistical approximation of the *possibility* of language itself.

The rise of LLM-authored texts also introduces epistemic and institutional dilemmas. In scientific publishing, for instance, most journals now require authors to disclose any use of generative AI. The prevailing trend goes further, forbidding the inclusion of LLMs as coauthors on the grounds that they cannot assume responsibility or consent to publication (*Nature*, 2023). These concerns are not merely bureaucratic: They reflect a deeper anxiety about the erosion of accountability, the shifting boundaries between drafting and delegation, and the difficulty of assessing provenance as machine-written and human-written texts are increasingly indistinguishable.

Moreover, as copyright becomes decoupled from creativity and authorship becomes distributed across systems, platforms, and user interactions, we may be witnessing the emergence of a new form of *algorithmic authorship* (Gretzky and Dishon, 2025). This is not a stable legal category but a sociotechnical assemblage: a hybrid of user intention, model output, and corporate infrastructure. It raises not only legal and economic concerns but ontological ones. If the author no longer stands behind the text, what does it mean to trust, interpret, or critique it?

Finally, the LLM reopens questions that Derrida (1976 [1967]) raises in *Of Grammatology* about writing as an autonomous, iterable system—detached from the presence of the speaker and resistant to closure. For Derrida, writing always exceeded intention and mastery; it was a form of 'trace' that eluded full recovery. LLMs radicalize this point: Their outputs are generated without experience, without presence, and without subjectivity. They are pure trace, unmoored from voice.

As we move into the next section, we will examine how this reconfiguration of authorship is already shaping concrete practices in knowledge production, from encyclopedic writing to academic publishing. There, the stakes become not only philosophical but institutional: How are norms, expectations, and responsibilities evolving as human and machine writing become increasingly entangled?

2 Empirical shifts: writing with and through large language models

While debates around authorship and originality provide an essential theoretical foundation, they take on new urgency when situated within empirical shifts in how text is actually produced across various domains. LLMs are no longer speculative tools— they are increasingly integrated into platforms of knowledge production, academic communication, and everyday expression. This section surveys three major contexts in which LLMs have already begun to alter writing practices: encyclopedic knowledge, scientific literature, and everyday digital genres.

Wikipedia and knowledge production. One of the most striking examples of LLM integration into public writing platforms is Wikipedia, a site historically built on community-driven, human-authored contributions. Recent work by Brooks et al. (2024) reveals that a significant number of edits to Wikipedia entries now originate from text generated with the help of language models, particularly in less-trafficked pages or non-English entries. The study, based on a large-scale analysis of revision histories and contributor patterns, highlights how contributors are increasingly turning to tools like ChatGPT to generate initial drafts, summarize sources, or rewrite existing content more fluently.

Importantly, the authors do not suggest that LLMs are being used covertly to deceive readers. Rather, their findings indicate that LLMs are increasingly employed to lower the cognitive and temporal barriers to participating in collaborative writing. These tools enable a wider range of users—including those with limited writing skills or without subject-matter expertise—to contribute to content creation.

However, this shift raises fundamental concerns about epistemic authority (Huang et al., 2025). Wikipedia operates under the principle of 'verifiability, not truth,' emphasizing that claims must be supported by reliable, published sources rather than personal knowledge or inference. This policy presupposes that contributors critically evaluate and cite authoritative sources. When contributors rely on LLM-generated text—often trained on the same public web content that Wikipedia itself draws upon—there is a risk of recursive citation (Algaba et al., 2025), in which unverifiable or weakly sourced claims are reproduced and reinforced rather than independently verified. Because LLMs are not infallible and can produce hallucinations—plausible but false or fabricated information—this process also increases the likelihood that factual errors will be introduced into Wikipedia, despite its editorial safeguards.

Moreover, as contributions increasingly originate from systems that obscure the line between user and tool, the traceability of authorship becomes more tenuous (see the previous section). This poses challenges to Wikipedia's foundational values of transparency, editorial accountability, and community-based governance.

Academic writing and scientific communication. In academic publishing, LLMs are also becoming increasingly common, not only as drafting tools but as silent coauthors. Geng and Trotta (2024) provide an analysis of LLM usage in scientific articles across disciplines. By means of a statistical analysis of word frequency changes, they show that a nontrivial proportion of submitted manuscripts now contain passages generated—at least in part—by systems such as ChatGPT or Claude. This is also the case for peer reviews, particularly those written by less confident reviewers or submitted at the last minute (Liang et al., 2024).

While many of these passages may have been checked and edited by the human author, the presence of LLM-generated segments introduces new risks. Among the most prominent are hallucinated citations and fabricated references—outputs that seem plausible but are in fact invented. These synthetic citations are difficult to detect through peer review, and their proliferation may contribute to epistemic noise in an already saturated scholarly environment.

Liang et al. (2024) also note that LLMs are used unevenly: Researchers from non-English-speaking backgrounds are more likely to rely on them for stylistic editing (as is the case for this book) or abstract generation. This suggests that LLMs may serve as equalizing tools in global academia, helping to mitigate linguistic inequalities. Yet the use of such tools remains stigmatized in many contexts, and most academic institutions lack clear norms or policies around disclosure and attribution.

Everyday writing: emails, reports, resumes. Beyond formal knowledge domains, LLMs are now woven into the fabric of everyday communication. Generative systems are becoming increasingly embedded in routine communicative practices, from Gmail's predictive text and GitHub Copilot's code suggestions to Microsoft's Office 365 Copilot. These tools now play a significant role in shaping the content of emails, job applications, and professional reports. Their adoption is not always visible—users may click 'autocomplete' without consciously recognizing that a neural model is shaping their phrasing.

Here again, these integrations raise questions about authorship, fluency, and linguistic agency. On one hand, LLMs improve accessibility: They assist users with limited language proficiency, help nonnative speakers formulate professional emails, and reduce the time spent on routine correspondence. On the other hand, they contribute to a flattening of style—a kind of algorithmic standardization that replaces voice with efficiency. The user becomes less of a writer and more of a curator, selecting among algorithmically suggested phrasings.

Here, the concern is less about intellectual property than about the erosion of expressive individuality. If writing becomes an act of prompt refinement rather than articulation, what happens to tone, nuance, and rhetorical development? Are we witnessing a subtle shift from composition to configuration—a deskilling of expressive labor in favor of convenience?

These developments call for a reconceptualization of authorship as a form of interaction shaped not solely by individual intention but by the affordances of digital platforms and the dynamics of model-mediated composition. This reconfiguration also complicates efforts to distinguish between human- and machine-generated text—a challenge we turn to in the next section.

3 Addressing the identification of AI-generated text

As LLMs become increasingly capable of producing coherent and stylistically convincing prose, concerns have mounted over how to reliably identify text generated by these systems. The capacity to attribute authorship and distinguish human- from machine-produced content is critical in many domains, including academic integrity, journalism, legal processes, and online content moderation.

These concerns reflect broader questions about accountability and transparency within the context of algorithmically mediated writing. In response, researchers and policymakers have developed various technical and procedural tools to address this challenge. Two prominent approaches are detection systems and watermarking techniques.

3.1 Detection of AI-generated text

As LLM-generated text becomes increasingly pervasive across domains, parallel efforts have emerged to develop tools capable of identifying it. These detectors are deployed in contexts ranging from academic assignments and scientific peer review to online disinformation and content moderation. Promoted as safeguards against misuse, detection systems are often presented as technical solutions to ethical problems. Yet empirical analysis reveals that this approach is deeply problematic, in terms of both its operational assumptions and its broader implications.

At the heart of the problem lies a basic definitional ambiguity. The distinction between 'human-written' and 'machine-generated' text is not a matter of content or form but of provenance—a property not accessible from the text itself. Most current detectors operate by analyzing surface linguistic features and statistical regularities. But these features do not reveal the generative process behind a text. Moreover, the very notion of 'LLM-generated' content is ill-defined. Detectors are typically trained on stylized benchmarks that involve unedited, zero-shot completions from a handful of models.

A troubling consequence of this simplification is the risk of false positives. No detection tool is infallible; most operate with error rates of 10–15%, and this rate increases under adversarial conditions or when dealing with short texts. Critically, these errors are not evenly distributed. Studies have shown that detection systems disproportionately misclassify content written by nonnative speakers or individuals with distinctive writing styles (Giray, 2024; Liang et al., 2023). In such cases, algorithmic judgment reinforces existing social inequities under the guise of technical neutrality. This echoes Ian Hacking's (1986) notion of 'looping effects,' where new classificatory schemes reshape the behavior and experiences of the people to whom they are applied. When those classifications are unreliable and socially consequential, they risk becoming not merely inaccurate but unjust.

Further complicating the issue is the role of postediting. In real-world usage, LLM outputs are rarely adopted verbatim. Users routinely revise, rearrange, or integrate generated text into longer human-authored documents. Yet most detectors are trained on raw completions, making them poorly suited to this hybrid reality. Even minimal human intervention—such as paraphrasing or synonym substitution—can drastically reduce detection accuracy. This fragility reveals the detectors' dependence on assumptions that no longer correspond to how LLMs are used in practice. Philosophically, this reflects a deeper misunderstanding of textuality. Attempts to reassign authorship based on stylometric features presuppose a transparency and traceability of origin that writing has never offered. As Derrida reminds us, text is always dispersed, iterated, and contextually shaped.

Another major concern is the ethical oversimplification introduced by binary classification. Ethical evaluation depends on context: Using an LLM to ghostwrite an academic article is not equivalent to using it for spelling correction or translation. Yet detection tools treat all forms of LLM assistance as equivalent. As Lepp and Smith (2025) and Cheng et al. (2025) argue, this erasure of nuance collapses a broad spectrum of human–AI interaction into a reductive dichotomy. The result is not just conceptual flattening but practical harm—making it difficult to distinguish between malicious automation and legitimate support.

Detection tools also face a serious generalization problem. Different LLMs—such as GPT-4, Claude, Gemini, and DeepSeek—produce outputs with distinct stylistic fingerprints. Tools trained on one model may underperform on another. Furthermore, newer LLMs increasingly incorporate alignment and posttraining procedures that make their outputs more human-like and therefore harder to detect (Antoun et al., 2024). Detectors thus lag behind the models they are meant to monitor, creating a moving-target problem. The situation is further exacerbated by the fact that real-world use often involves prompt engineering, in-context learning, or style adaptation—practices that push generated text further outside the training distribution of most detectors.

Finally, detection tools are rarely interpretable. Most function as black boxes, offering no clear rationale for their decisions. This lack of transparency undermines their trustworthiness—especially when used in high-stakes contexts like academic integrity enforcement, scientific publication, or employment screening. Lipton (2018) emphasizes that opacity limits both the legitimacy and the accountability of automatic systems. Yet, in public discourse, such systems are often misunderstood as definitive arbiters of authorship. In reality, they are statistical guessers with well-documented biases and limitations.

Ultimately, we can say that the current generation of LLM detection tools offers, at best, a partial and fragile solution to a complex sociotechnical problem. Rather than resolving concerns about AI misuse, they risk amplifying inequalities, institutionalizing suspicion, and reinforcing reductive understandings of writing and authorship. As we will see in Section 4, these tensions are especially visible—and especially consequential—in the field of education.

3.2 Text watermarking

A complementary strategy is text watermarking, which involves embedding subtle, often imperceptible signals within a language model's output to mark it as synthetic. Unlike detection, which occurs after the text is produced, watermarking proactively labels content at the moment of generation, enabling later verification of its origin.

In high-stakes contexts—such as governmental communication, educational assessment, or scientific publishing—watermarking can help establish provenance and support trust. Some researchers argue that robust watermarking techniques could be designed to withstand moderate paraphrasing or editing, making them a reliable tool if widely adopted and standardized (Kirchenbauer et al., 2023). However, watermarking also faces significant limitations. Not all generative models will implement watermarking protocols, leaving open the possibility of unmarked synthetic text. Additionally, texts can be heavily edited, reformatted, or translated in ways that disrupt or erase embedded signals, undermining watermark effectiveness in practice.

Moreover, broad adoption of watermarking across diverse commercial and open-source systems presents major technical and political challenges. There are also concerns about potential negative consequences for user privacy and freedom of expression, especially if watermarking mechanisms enable intrusive forms of traceability (Fernandez et al., 2024).

Consequently, while watermarking offers a promising complement to detection methods, it cannot provide a comprehensive safeguard. Its usefulness depends on coordinated adoption, robust technical standards, and integration with broader policy and educational frameworks, alongside careful consideration of ethical and civil liberties implications.

4 Education and the future of writing

The integration of LLMs into educational settings has sparked vigorous debate, revealing both hopes and anxieties about the future of writing, learning, and intellectual development. At stake is not merely the question of academic honesty but a deeper set of concerns about the nature of cognitive effort, the function of writing in the development of thought, and the evolving role of the teacher in a technologically mediated environment. This section examines three dimensions of this debate: the risk of deskilling and the perceived erosion of originality; the potential of LLMs to foster greater equity and inclusion; and the need to reconceive pedagogical approaches in light of these new tools.

4.1 The risk of deskilling and the crisis of originality

Perhaps the most frequently expressed concern among educators is that LLMs may encourage a form of intellectual outsourcing. Students, it is feared, may come to rely on generative systems not only to polish their prose but to compose it entirely—bypassing the cognitive labor involved in formulating arguments, organizing ideas, and experimenting with voice and style (Cotton et al., 2023). Writing, in this view, is not merely a vehicle for displaying knowledge but a process through which understanding is constructed. If that process is delegated to machines, the formative function of writing may be compromised.

These concerns are often framed in terms of 'deskilling,' a concept borrowed from labor studies to describe the erosion of expertise through technological automation (Crowston and Bolici, 2024). Just as assembly-line technologies once displaced artisanal labor, so too might LLMs displace core academic competencies. This worry resonates with earlier critiques of spell-checkers and grammar-correction software but it is amplified by the scope and fluency of contemporary language models, which can now produce extended, contextually appropriate texts that closely approximate the style and conventions of academic and professional writing.

In philosophical terms, this anxiety reflects a long-standing association between authorship and agency. Writing has historically been seen as a mark of autonomy, creativity, and intellectual responsibility. If students submit texts they have not composed, they risk alienating themselves from their own ideas. Hannah

Arendt (1958) warns that modernity risks dissolving the 'space of appearance' in which individuals reveal themselves through word and deed—a space that, in educational contexts, includes the authorship of one's own writing. The fear, then, is not simply that students will cheat but that they will cease to think—and cease to appear—as authors.

4.2 Opportunity for equity and inclusion

Yet this line of critique, while important, can obscure the considerable benefits that LLMs offer to many students—particularly those for whom academic writing presents disproportionate challenges. For learners with limited proficiency in the language of instruction, or those with dyslexia, cognitive impairments, or anxiety disorders, generative models can act as enablers rather than crutches (Goodman et al., 2022). They can provide lexical alternatives, suggest syntactic rephrasings, or offer structural scaffolding that facilitates engagement rather than replaces it.

In multilingual or under-resourced educational contexts, where instructional support may be minimal and feedback scarce, access to language models can help mitigate structural inequalities. Rather than eroding the quality of writing, these systems may allow students to participate in discursive practices from which they would otherwise be excluded. The emphasis on linguistic correctness, often wielded as a gatekeeping device in academic institutions, can be softened by tools that assist rather than penalize.

This point has been raised by scholars studying global academic publishing. Amano et al. (2023) argue that nonnative English speakers face systemic disadvantages in science communication, often expending more time and labor to produce work that is judged on stylistic grounds. LLMs, if used transparently and ethically, may help redress such imbalances by reducing the cognitive and linguistic load required to meet normative standards.

At a deeper level, this raises the question of what counts as legitimate support. If a student uses a thesaurus or a grammar guide, their work is not typically disqualified. LLMs challenge the boundary between permissible aid and disqualifying automation, forcing educators to reconsider the principles that underlie their assessments.

4.3 Reinventing pedagogy

These shifts suggest that educational institutions must not simply prohibit or ignore the presence of LLMs, but reimagine pedagogy in their wake (Holmes et al., 2019). One promising direction involves moving from the evaluation of final products to the assessment of processes. Rather than judging the end result alone, instructors might emphasize iterative writing, reflection, peer review, and the documentation of revision histories. This would make visible the cognitive labor involved in composition, regardless of whether LLMs are used.

Another strategy lies in cultivating what might be called 'tool literacy.' Instead of banning generative models, educators could train students to understand their affordances and limitations, to critically evaluate their outputs, and to use them responsibly. This aligns with a broader pedagogical shift from content mastery to metacognitive awareness: knowing how to learn, how to write, and how to engage with evolving technologies in an informed and reflective manner.

Transparency is key. Students should be encouraged—and in some cases required—to disclose their use of LLMs, specifying what parts of their text were assisted or generated, and why. Such practices not only uphold academic integrity but model a form of accountable collaboration between human and machine. Importantly, this does not mean treating LLMs as coauthors but rather situating them as instruments—whose use must be acknowledged, contextualized, and justified.

More broadly, these developments invite a more pluralistic conception of writing. Not all writing must be original in the same way; not all uses of assistance are illegitimate. The goal is not to preserve a narrow ideal of the autonomous writer but to establish adaptable standards that foster meaningful intellectual engagement and equitable participation within a writing environment shaped by both human and machine contributions.

5 Conclusion: writing after writing

LLMs do not simply assist in writing—they transform its conditions of possibility. Like earlier technologies such as the printing press or the word processor, they alter the means and modes of textual production. But unlike those tools, which extended human capacities without displacing human agency, LLMs are capable of generating entire discourses: essays, arguments, summaries, and narratives, often indistinguishable from those composed by human authors. As a result, they blur foundational distinctions between assistance and authorship, originality and reproduction, intention and automation.

This transformation has significant implications for our understanding of writing, authorship, and intellectual responsibility. As the production of texts becomes increasingly mediated by algorithmic systems, questions of provenance—who authored a given text, under what conditions, and with what epistemic or institutional commitments—become more complex and, at times, opaque. The central concern is not the hypothetical emergence of artificial general intelligence but a more immediate and concrete issue: the weakening of established frameworks for attributing meaning, accountability, and authority in textual practices.

In this context, the challenge is to reassess and rearticulate the norms and infrastructures that support scholarly communication, educational evaluation, and public discourse. Rather than framing the rise of generative models as a disruption to be resisted, it should be understood as a shift that demands new critical, institutional, and conceptual tools—capable of responding to the reconfiguration of authorship and knowledge production in an increasingly algorithmically mediated world.

References

Algaba, A., Mazijn, C., Holst, V., Tori, F., Wenmackers, S., and Ginis, V. (2025). Large language models reflect human citation patterns with a heightened citation bias. In *Findings of the Association for Computational Linguistics (NAACL 2025), Albuquerque, USA (pp. 6829–6864).* Association for Computational Linguistics. https://aclanthology .org/2025.findings-naacl.381/

Amano, T., González-Varo, J. P., and Sutherland, W. J. (2023). Languages are still a major barrier to global science. *PLOS Biology, 21*(4). https:// doi.org/10.1371/journal.pbio.2000933

Antoun, W., Sagot, B., and Seddah, D. (2024). From text to source: Results in detecting large language model-generated content. In *Proceedings of the 2024 Joint International Conference on Computational Linguistics, Language Resources and Evaluation (LREC-COLING 2024), Turin, Italy* (pp. 7531–7543). ELRA and ICCL. https://aclanthology.org/2024.lrec-main.665.pdf

Arendt, H. (1958). *The human condition.* University of Chicago Press.

Benjamin, W. (1969 [1935]). *The work of art in the age of mechanical reproduction.* In: H. Arendt (ed.), *Illuminations* (H. Zohn, trans.), Schocken Books.

Brooks, C., Eggert, S., and Peskoff, D. (2024). The rise of AI-generated content in Wikipedia. In *Proceedings of the First Workshop on Advancing Natural Language Processing for Wikipedia, Miami, USA* (pp. 67–79). Association for Computational Linguistics. https:// aclanthology.org/2024.wikinlp-1.12/

Cheng, Z., Zhou, L., Jiang, F., Wang, B. and Li, H. (2025). Beyond binary: Towards fine-grained LLM-generated text detection via role recognition and involvement measurement. In *Proceedings of the ACM on Web Conference 2025* (pp. 2677–2688). Association for Computing Machinery. https://dl.acm.org/doi/10.1145/3696410.3714770

Cotton, D. R. E., Cotton, P. A., and Shipway, J. R. (2023). Chatting and cheating: Ensuring academic integrity in the era of ChatGPT. *Innovations in Education and Teaching International, 61*(2), 228–239. https://doi.org/10.1080/14703297.2023.2190148

Crowston, K., and Bolici, F. (2024). Deskilling and upskilling with generative AI systems. *Information Research, 30,* 1009–1030.

Derrida, J. (1976 [1967]). *Of grammatology* (G. C. Spivak, Trans.). Johns Hopkins University Press.

Fernandez, P., Level, A., and Furon, T. (2024). What lies ahead for generative AI watermarking. In *ICML 2024: 2nd Workshop on Generative AI and Law (GenLaw '24), Vienna, Austria*. ICML. https://icml.cc/virtual/2024/39181

Foucault, M. [1977 (1969)]. What is an author? In D. F. Bouchard and S. Simon (Trans. and Eds.), *Language, counter-memory, practice* (pp. 113–138). Cornell University Press.

Geng, M., and Trotta, R. (2024). Is ChatGPT transforming academics' writing style? In *ICML 2024 Workshop: Next Generation of AI Safety, Vienna, Austria*. ICML. https://arxiv.org/abs/2404.08627

Goodman, S. M., Buehler, E., Clary, P., Coenen, A., Donsbach, A., Horne, T. N., Lahav, M., MacDonald, R., Michaels, R. B., Narayanan, A., Pushkarna, M., Riley, J., Santana, A., Shi, L., Sweeney, R., Weaver, P., Yuan, A., and Ringel Morris, M. (2022). LaMPost: Design and evaluation of an AI-assisted email writing prototype for adults with dyslexia. In *Proceedings of the 24th International ACM SIGACCESS Conference on Computers and Accessibility (ASSETS '22)* (pp. 1–18). Association for Computing Machinery. https://doi.org/10.1145/3517428.3544819

Giray, L. (2024). The problem with false positives: AI detection unfairly accuses scholars of AI plagiarism. *The Serials Librarian, 85*(5–6), 181–189.

Gretzky, M., and Dishon, G. (2025). Algorithmic-authors in academia: Blurring the boundaries of human and machine knowledge production. *Learning, Media and Technology*. preprint. https://doi.org/10.1080/17439884.2025.2452196

Hacking, I. (1986). Making up people. In T. C. Heller, M. Sosna, and D. E. Wellbery (Eds.), *Reconstructing individualism: Autonomy, individuality, and the self in Western thought* (pp. 222–236). Stanford University Press.

Holmes, W., Bialik, M., and Fadel, C. (2019). *Artificial intelligence in education: Promises and implications for teaching and learning*. Center for Curriculum Redesign.

Huang, S., Xu, Y., Geng, M., Wan, Y., and Chen, D. (2025). Wikipedia in the era of LLMs: Evolution and risks. arXiv preprint. https://arxiv.org/abs/2503.02879

Kirchenbauer, J., Geiping, J., Wen, Y., Katz, J., Miers, I., and Goldstein, T. (2023). A watermark for large language models. In *Proceedings of the 40th International Conference on Machine Learning, PMLR, 202,*

17061–17084. https://proceedings.mlr.press/v202/kirchenbauer23a .html

Lepp, H. and Smith, D. S. (2025). 'You cannot sound like GPT': Signs of language discrimination and resistance in computer science publishing. In *Proceedings of the ACM Conference on Fairness, Accountability, and Transparency (FAccT), Athens, Greece.* Association for Computing Machinery. https://dl.acm.org/doi/10.1145/3715275 .3732202

Liang, W., Izzo, Z., Zhang, Y., Lepp, H., Cao, H., Zhao, X., Chen, L., Ye, H., Liu, S., Huang, Z., McFarland, D., and Zou, J. (2024). Monitoring AI-modified content at scale: A case study on the impact of Chat-GPT on AI conference peer reviews. In *International Conference on Machine Learning (ICML), Vienna, Austria.* Association for Computing Machinery. https://dl.acm.org/doi/10.5555/3692070.3693262

Liang, W., Yuksekgonul, M., Mao, Y., Wu, E., and Zou, J. (2023). GPT detectors are biased against non-native English writers. *Patterns, 4*(7), 100779.

Lipton, Z. C. (2018). The mythos of model interpretability: In machine learning, the concept of interpretability is both important and slippery. *ACM Queue, 16*(3), 31–57. https://doi.org/10.1145/3236386.3241340

Nature (2023). Editorial: Tools such as ChatGPT threaten transparent science; here are our ground rules for their use. *Nature, 613*(7945), 612. https://doi.org/10.1038/d41586-023-00191-1

Underwood, T. (2023). The empirical triumph of theory. *Critical Inquiry* [Blog]. https://critinq.wordpress.com/2023/06/29/the-empirical-tri umph-of-theory/

US Copyright Office. (2023). Copyright registration guidance: Works containing material generated by artificial intelligence. *Federal Register, 88*(52), 16190–16194.

PART II

The risks of anthropomorphism (large language models and the mind)

CHAPTER 4

Large language models and reasoning, the boundaries of mind and consciousness

Large language models (LLMs) have quickly become central figures in contemporary discussions about artificial intelligence, cognition, and even the nature of thought itself (Millière and Buckner, 2024a, 2024b). As we saw in Chapter 1, these models, trained on vast datasets, detect statistical regularities and then generate coherent and contextually appropriate continuations of text. They are, at their core, probabilistic engines, without any explicit understanding of the world they describe. LLMs possess no formal model of reality, no sensory grounding, no body through which to act or perceive. Yet their outputs often display a level of fluency, coherence, and flexibility that invites comparison to human cognition.

What is striking is how readily people interpret the behavior of these models in cognitive and psychological terms. It has become common to speak of LLMs as if they 'understand,' 'reason,' or 'imagine.' Researchers evaluate their performance on tasks designed to test commonsense knowledge, logical reasoning, or theory of mind. Public discourse is rife with speculation about

How to cite this book chapter:
Poibeau, T. 2025. *Understanding Conversational AI: Philosophy, Ethics and Social Impact of Large Language Models.* Pp. 85–112. London: Ubiquity Press. DOI: https://doi.org/10.5334/bde.e. License: CC BY-NC 4.0

whether such models are, or could become, conscious. In short, there is a persistent tendency to map human cognitive categories onto systems that, by design, lack the biological and phenomenological substrates typically associated with those capacities.

This projection is not entirely surprising. Language itself is a primary vehicle of human cognition, a mirror of mental life. When an artificial system wields language with proficiency, it becomes natural to ascribe to it a mind-like quality. Moreover, LLMs can successfully imitate behaviors associated with cognitive abilities: They can answer complex questions, generate novel ideas, and even simulate the reasoning patterns of human agents. Yet this raises a critical question: Are these ascriptions merely metaphorical, useful heuristics for human observers? Or do they point to a functional equivalence that blurs the distinction between biological and artificial cognition?

At the heart of this inquiry lies a deeper philosophical issue. Do the capabilities of LLMs reveal something fundamental about the nature of cognition itself? Perhaps human cognitive processes are, to a significant degree, statistical and pattern-based, more similar to the operations of these models than traditionally assumed. Alternatively, the apparent similarities may mask profound differences—differences rooted in embodiment, intentionality, or consciousness—that no amount of linguistic prowess can bridge. In this chapter, we return to this tension: whether LLMs genuinely partake in cognitive processes or whether they are sophisticated simulacra, reflecting the surface features of thought without participating in its deeper structures.

This chapter explores what it means to say that language models 'understand,' 'reason,' or even appear to 'possess minds.' As these systems increasingly mimic behaviors associated with human intelligence, they raise urgent questions about the boundaries between simulation and cognition, pattern recognition and understanding, language fluency and conscious thought. We examine how LLMs handle core aspects of cognition—such as commonsense reasoning, inferential logic, and the ability to attribute mental states—and ask whether these capacities reflect genuine cognitive processes or merely the appearance of such. By drawing on insights from cognitive science, philosophy of mind, and AI research, we aim to understand not only what LLMs can do but what their performance reveals about the nature of cognition itself—and what might still set human thought apart.

1 Large language models and common sense: statistical knowledge versus world knowledge

Common sense has traditionally been regarded as one of the most elusive and foundational components of human cognition. It encompasses an extensive, often implicit body of world knowledge—facts about physical objects, social norms, causal regularities, and everyday expectations—that enables humans to navigate novel situations flexibly and efficiently. Philosophers from Aristotle to Reid (2022 [1764]) have emphasized its essential role in practical reasoning, while contemporary cognitive science explores related foundational cognitive structures known as core knowledge. Core knowledge theories propose innate, domain-specific systems that underlie humans' intuitive understanding of objects, actions, numbers, space, and social interactions (Carey, 2009; Spelke and Kinzler, 2007).

The question of whether machines, particularly LLMs, possess common sense has garnered increasing attention. On the surface, LLMs display an impressive capacity to produce responses that align with commonsensical knowledge. Benchmarks such as the CommonSenseQA dataset (Talmor et al., 2019) or the Winograd Schema Challenge (Levesque et al., 2012) demonstrate that modern LLMs can achieve performance levels rivaling or surpassing human baselines, especially when fine-tuned or prompted effectively.

For instance, a typical question from the CommonSenseQA dataset might ask, 'Where on a river can you hold a cup upright to catch water on a sunny day? (a) waterfall, (b) bridge, (c) valley, (d) pebble, (e) mountain.' This question assesses the model's capability to leverage commonsense knowledge about typical object locations, with the correct answer being (a), waterfall. The Winograd Schema Challenge presents a different kind of commonsense reasoning problem. An example of such a schema is: 'The trophy did not fit into the suitcase because it was too large. What was too large, the trophy or the suitcase?' Here, the correct interpretation—that 'it' refers to the trophy—requires an understanding of everyday physics and practical reasoning. Both these examples illustrate the nuances involved in evaluating the extent to which LLMs truly grasp or merely simulate commonsense understanding.

As we saw in the previous chapters, this apparent mastery of common sense is in fact largely statistical in nature. LLMs internalize cooccurrence patterns and conditional probabilities across massive corpora of text, allowing them to produce answers that seem consistent with everyday knowledge. Their 'knowledge' is not grounded in an embodied experience of the world but extracted from textual regularities—a point that differentiates them starkly from human cognitive agents. Moreover, many benchmarks used to evaluate commonsense reasoning are publicly available online or included in training data. This means that LLMs often achieve high performance not because they truly 'understand' the tasks but because they have already seen similar or even identical examples during training. Consequently, benchmark performance is not always a reliable indicator of genuine reasoning capabilities.

One way to address this problem is to ensure that evaluation datasets are carefully curated to avoid overlap with training data—either by keeping benchmarks private until evaluation or by using held-out, proprietary datasets. For example, ARC-Challenge (Chollet, 2019) and BIG-bench (Srivastava et al., 2022) have taken steps to track data provenance or release subsets after model training.

Another promising technique involves counterfactual test sets designed to test whether a model can generalize beyond surface patterns by breaking spurious correlations (Chen et al., 2025). In the domain of commonsense reasoning, this might involve altering a scenario to test whether the model relies on actual causal or physical understanding rather than memorized associations. For example, a model might be trained to answer:

'If you drop a glass on the floor, what is likely to happen?'—with the expected answer being 'It breaks.'

A counterfactual version might ask:

'If you drop a glass on a thick carpet, what is likely to happen?'—where the correct answer might be 'It might not break.'

A model that has only learned the association 'glass → break' may fail to take the context into account, whereas one demonstrating deeper reasoning would adjust its answer appropriately (but determining the correct answer in this context is then not immediately obvious, even for humans). Such tests are crucial for evaluating whether a model's apparent competence reflects true generalization or shallow pattern recognition.

This reliance on surface-level associations—rather than grounded understanding—has led to a broader critique within AI and philosophy: that language models, despite their fluency, lack genuine understanding of the world they describe. Hubert Dreyfus (1972), a prominent critic of symbolic AI, argues that intelligence cannot be reduced to formal rule-based manipulation detached from embodied experience. Although LLMs do not rely on explicit symbolic rules, their statistical approach remains similarly disembodied. Their 'understanding' of objects, actions, and causes is filtered entirely through textual patterns, without the sensorimotor grounding that underlies human cognition. This lack of grounding tends to become most apparent when modern LLMs encounter highly implausible scenarios or tasks requiring causal or temporal reasoning. For instance, Kevin Lacker reports that GPT-3 when asked 'How many eyes does a blade of grass have?' confidently replied 'A blade of grass has one eye,'[1] rather than recognizing the absurdity of the question. Such errors reveal limitations beneath their generally

---

[1] Lacker, K. (2020, July). Giving GPT-3 a Turing Test [Blog post]. Retrieved June 29, 2025, from https://lacker.io/ai/2020/07/06/giving -gpt-3-a-turing-test.html.

robust surface performance—though this issue is less frequent in more recent models.

Cognitive scientists like Spelke and Kinzler (2007) emphasize that human infants develop core knowledge systems through embodied interaction with the world. LLMs, by contrast, lack any developmental trajectory grounded in perception and action as noted above. Still, it is worth acknowledging that textual corpora do encode substantial amounts of implicit world knowledge. Linguistic structures such as enumerations ('animals such as cats, dogs, and rabbits') convey taxonomic and relational information that LLMs can use to simulate an understanding of categories and associations (Hearst, 1992). This simulation is not equivalent to human common sense, but it does allow for a form of pragmatic competence grounded in language use.

The question of how machines might achieve a more robust and genuine form of commonsense reasoning has been a central concern in AI since its inception. John McCarthy (1979), one of the field's founders, argues that AI urgently needs a 'science of common sense.' Such a science, he proposes, should address how knowledge is acquired, accessed, and represented in ways that connect meaningfully to the world. McCarthy emphasizes the importance of distinguishing between different types of representations—those derived from direct experience versus those abstracted through formal reasoning—and understanding how each contributes to intelligent behavior. As discussed in Chapter 1, these challenges are closely linked to the problem of grounding—how symbols and representations in artificial systems can meaningfully connect to perception, action, and the physical world. Without such grounding, even the most elaborate formal systems risk remaining unmoored from the kinds of intuitive, embodied understanding that characterize human cognition.

One of the most ambitious responses to McCarthy's call emerged in the 1980s with the work of Doug Lenat, who launched the Cyc project as a large-scale attempt to build a formalized repository of common sense (Lenat and Guha, 1990). Cyc aimed to manually encode vast amounts of everyday knowledge into a structured database of logical assertions, hoping to capture the inferential richness of ordinary reasoning. However, despite decades of painstaking effort, the project fell short of delivering on its ambitious goals. One significant challenge it encountered was the extreme context-dependence of common sense, where the same piece of knowledge could radically shift in meaning based on subtle contextual cues—a flexibility that humans navigate effortlessly but that resists rigid formalization. As discussed in Chapter 2, this issue closely parallels the difficulties of formalizing natural language, which is similarly shaped by context, pragmatics, and implicit background assumptions.

Ultimately, while LLMs demonstrate a compelling mimicry of common sense grounded in statistical associations, true commonsense understanding remains elusive, embedded as it is in the lived, perceptual, and context-sensitive fabric of human experience. Yet textual data still offers vast reservoirs of commonsense knowledge drawn from cultural and social practices, which LLMs can partially exploit. Multimodal models that also integrate visual or video data—showing how objects and events unfold in the real world—may add a richer source of grounding, especially for reasoning about everyday physics and causal interactions. Even without a formal, systematic model of common sense, these systems can nonetheless master a surprisingly large amount of practically useful information, helping to explain why the frequency of nonsensical outputs has dropped significantly in just a few years.

2 Large language models and reasoning: probabilistic inference versus symbolic logic

Reasoning has long been conceived, in both philosophy and cognitive science, as a process of symbolic manipulation governed by formal rules of logic. From Aristotle's syllogisms to Frege's predicate calculus, logical reasoning is often framed as the application of systematic, explicit principles to derive conclusions from premises. Cognitive models such as the mental logic hypothesis

(Rips, 1994) and the mental models theory (Johnson-Laird, 1983) posit that human reasoning operates via internal representations that mirror these formal systems. Yet, with the advent of LLMs, a new, fundamentally different approach to reasoning has emerged—one that eschews symbolic manipulation in favor of probabilistic inference over vast linguistic corpora.

LLMs, based on deep neural architectures, do not encode logical rules explicitly. Instead, they capture statistical patterns within language data, enabling them to generate plausible continuations of text based on context. Remarkably, despite lacking explicit logical mechanisms, these models exhibit emergent reasoning-like abilities. Recent studies have demonstrated that, when prompted appropriately, LLMs can engage in tasks traditionally associated with reasoning: solving arithmetic word problems, following chains of reasoning in chain-of-thought prompting (Wei et al., 2022), performing analogical reasoning (Webb et al., 2023), and adapting behavior through in-context learning.

The integration of programming code and mathematical problem-solving exercises into the training data of LLMs has significantly enhanced their grasp of logical structures. By learning from structured datasets containing explicit logical and algorithmic steps, these models implicitly internalize patterns corresponding to reasoning processes. For instance, OpenAI's GPT models, trained extensively on code repositories, demonstrate an impressive capacity to produce coherent algorithms, debug snippets, and even suggest logical improvements to existing code. Such performance illustrates their capability to represent and operationalize structured, step-by-step reasoning processes, at least implicitly.

But do LLMs genuinely reason, or do they merely reproduce linguistic patterns that correlate with reasoning? Philosophically, this evokes questions about the nature of underlying cognitive capacities. While Chomsky's distinction between competence and performance (Chomsky, 1965) emphasizes a deep internal system governing linguistic abilities, applying this framework to LLMs is problematic. Unlike human speakers, whose competence reflects systematic and generative rule-based knowledge,

LLMs lack an explicit, introspectively accessible system of reasoning rules. Their apparent reasoning is instead emergent from statistical associations in the training data. Critics argue that this results in a form of reasoning that is epiphenomenal—an artifact of surface-level pattern-matching rather than the outcome of genuine inferential processes.

The cognitive scientist Daniel Kahneman's dual-system theory (Kahneman, 2011) offers a useful lens for analyzing this issue. According to Kahneman, human cognition operates via two systems: System 1, which is fast, automatic, and intuitive, and System 2, which is slow, deliberate, and reflective. LLMs excel at tasks reminiscent of System 1—generating fluent language, completing patterns, and making quick associations. However, they struggle with tasks that demand sustained, reflective, and metacognitive engagement characteristic of System 2 reasoning. Their lack of explicit representations and rule-based architectures inhibits their ability to engage in rigorous, stepwise logical deduction beyond the superficial simulation of such behavior (however, this is already developing quickly).

The limitations of LLMs' reasoning capabilities become evident in their susceptibility to hallucinations—producing outputs that are syntactically coherent but factually or logically incorrect. Without explicit mechanisms to check consistency, validity, or truth, LLMs often fail to detect contradictions or logical fallacies within their outputs. This further highlights their reliance on probabilistic associations rather than formal reasoning. It is also worth noting that LLMs tend to perform worse on reasoning tasks involving rarer tokens, even when the task itself is structurally identical, highlighting their dependence on frequency-driven patterns in training data.

Recent research by Shojaee et al. (2025) further reinforces these concerns by evaluating large reasoning models (LRMs), which explicitly aim to extend reasoning capabilities beyond those of standard LLMs. Their study demonstrates that, while LRMs initially show improved performance on tasks of moderate complexity, they experience a complete accuracy collapse as problem complexity increases beyond certain thresholds. Even with ample inference budgets, these models paradoxically reduce their reasoning effort near the collapse point, revealing a fundamental limitation in scaling reasoning abilities. Moreover, Shojaee et al. identify distinct performance regimes: standard LLMs are often more efficient on simpler tasks, while LRMs excel on medium-complexity problems thanks to chain-of-thought mechanisms, yet both architectures fail completely on highly complex, generalizable reasoning challenges.

Perhaps even more striking is the finding that LRMs struggle with precise, step-by-step logical execution, performing poorly even when explicitly provided with correct algorithms—for example, failing to solve classic problems such as the Tower of Hanoi once complexity increases. Analysis of their internal reasoning traces further reveals inefficient or counterproductive 'thinking' patterns, such as overexploring incorrect solutions on simple problems, or failing entirely to converge on correct answers for difficult ones. Shojaee et al.'s carefully controlled experiments in classic planning domains provide strong evidence that current LRMs' apparent 'thinking' abilities are often an illusion, limited to well-studied benchmarks and breaking down under more systematically varied, controlled, and genuinely challenging conditions. Together, these observations highlight that, despite their impressive surface-level performance, even advanced reasoning models remain constrained by data-driven pattern-matching and lack the robust, generalizable reasoning competencies required for truly systematic logical thought.

Additionally, LLMs lack what is traditionally considered metacognitive awareness. Human reasoners can reflect on their reasoning processes, recognize errors, and revise their inferences accordingly. In contrast, current LLMs do not possess explicit self-monitoring mechanisms or an independent capacity for error correction beyond what is implicitly captured in their training data. Philosophers such as Robert Brandom (1994) have argued that reasoning is not merely a matter of generating correct inferences but one of participating in a normative space of giving and asking for reasons—what he terms the 'game of giving and asking for reasons' (GOGAR). Within this framework, reasoning is fundamentally dialogical and normative: Any participant can challenge or justify claims to maintain coherence with other commitments. This dimension of explicit justification and reciprocal challenge is still largely absent from LLM behavior.

However, recent research suggests a more nuanced picture. For example, Ferrando et al. (2025) show that LLMs may encode a form of 'knowledge awareness' in their internal representations, identifying whether they recognize a given entity or not. Using sparse autoencoders, they find evidence that LLMs can steer their outputs depending on whether an entity is known, suggesting something akin to self-knowledge about their own limitations. This form of implicit metacognition does not amount to full reflective awareness but indicates that models may contain internal structures that track the boundaries of their own knowledge, influencing whether they hallucinate or refuse to answer.

Furthermore, developments in alignment techniques have explored how models might acquire a more normative, Brandom-style capacity to self-critique. Anthropic's Constitutional AI approach (Bai et al., 2022), for example, aims to train models to revise their own responses based on a constitution-like principles document. In this setting, the model is guided to critique its outputs according to a set of normative constraints, potentially fostering greater consistency and coherence over time. In parallel, other approaches—such as DeepSeek's chain-of-thought prompting combined with reinforcement learning—attempt to scaffold explicit reasoning by encouraging models to elaborate their intermediate steps and then reinforce those explanations.

While these methods fall short of instilling genuine metacognitive subjectivity, they nonetheless nudge LLMs closer to explicit reasoning processes and norm-based evaluation. Although LLMs still lack a fully developed normative reasoning faculty comparable to human metacognition, emerging research points to ways in which limited forms of knowledge awareness and norm-constrained self-monitoring can be partially encoded within their architectures. These directions hold promise for reducing hallucinations and improving trustworthiness, although they do not yet amount to a true 'science of common sense' in McCarthy's original sense.

3 Large language models and theory of mind: simulation without mentalization

One of the central constructs in cognitive psychology is theory of mind (ToM): the capacity to attribute mental states—beliefs, desires, intentions, knowledge—to oneself and to others. This ability allows humans to predict, interpret, and respond to the behavior of others in social contexts. Developmental psychologists, following the foundational work of Premack and Woodruff (1978), have extensively studied how ToM emerges in childhood, with particular focus on tasks such as the false-belief test, where success depends on understanding that another individual can hold a belief different from reality and from one's own knowledge.

Recent empirical research has shown that certain LLMs, particularly the largest ones, can pass simplified versions of these same ToM tasks (Kosinski, 2024). For instance, Kosinski tested GPT-3.5 and GPT-4 on classic false-belief scenarios such as the Sally-Anne task: 'Sally places a marble in a basket and leaves the room; while she is away, Anne moves the marble to a box.' The question posed to the model is: 'Where will Sally look for the marble when she returns?' Remarkably, when prompted appropriately, the LLMs correctly answer that Sally will look in the basket, demonstrating an ability to predict that Sally will act based on her outdated belief, not the actual state of the world. This raises an immediate and intriguing question: Do LLMs genuinely possess something akin to ToM, or are they merely simulating the linguistic patterns associated with it, without any underlying understanding?

It is generally assumed that, despite their impressive behavioral success in many tasks, LLMs do not construct internal models of others' beliefs or intentions in the way humans do. Their outputs are typically guided by statistical regularities in the training data, rather than by a genuine attribution of mental states. In this sense, they appear to engage in pattern recognition without true mentalization, lacking access to a self-representation or to stable representations of distinct agents with their own perspectives. Recently, Anthropic claimed to have identified circuits within LLMs that seem to track beliefs about other agents across narratives and dialogues, hinting at the emergence of primitive, distributed forms of perspective-taking (Ameisen et al., 2025; Lindsey et al., 2025). While these representations do not amount to fully human-like ToM, they suggest that LLMs may develop partial, functionally relevant analogues through large-scale training.

Philosophically, this raises the issue of functional equivalence versus genuine cognition. Is the outward display of ToM-like behavior sufficient to attribute ToM capabilities or does genuine understanding require more? Daniel Dennett's notion of the intentional stance (Dennett, 1987) is instructive here. Dennett argues that we often explain the behavior of complex systems—whether humans, animals, or even thermostats—by attributing beliefs and desires, regardless of whether these systems possess actual mental states.

Here is how it works: First you decide to treat the object whose behavior is to be predicted as a rational agent, then you figure out what beliefs that agent ought to have, given its place in the world and its purpose. Then you figure out what desires it ought to have, on the same considerations, and finally you predict that this rational agent will act to further its goals in the light of its beliefs. A little practical reasoning from the chosen set of beliefs and desires will in most instances yield a decision about what the agent ought to do; that is what you predict the agent will do.

In this framework, adopting the intentional stance means attributing beliefs, desires, and rationality to a system not because it necessarily possesses these properties intrinsically, but because doing so offers a fruitful strategy for predicting its behavior. Applied to artificial agents like LLMs, this approach does not claim that the model has an inner mental life but rather that treating it *as if* it does can be pragmatically useful. From this perspective, if treating an LLM as if it has beliefs and intentions yields successful predictions, there may be pragmatic value in adopting the intentional stance toward it.

However, many would argue that functional success is not enough. The intentional stance is, after all, a heuristic. Without real mental states—without subjective experience, internal perspectives, or recursive self-awareness—the model lacks what cognitive scientists typically consider essential to ToM. Unlike a child or an adult human, an LLM is generally taken to lack genuine knowledge of its own knowledge; it does not appear to reflect on its representations or adjust its behavior based on an internal model of others' mental states. Still, it cannot be entirely ruled out that some latent representations might support a limited or implicit form of self-monitoring or recursive processing, though current evidence does not definitively establish this. This absence—or at least extreme minimality—of recursive mentalizing and self-other differentiation highlights a key limitation.

Furthermore, LLMs suffer from brittleness in ToM tasks. Slight alterations in phrasing, context, or complexity often lead to failure, revealing the absence of robust, underlying cognitive mechanisms (Hu et al., 2025; Ullman, 2023). Studies have also shown that LLMs do not generalize ToM capabilities across novel or dynamically changing scenarios—something humans do effortlessly.

An additional limitation is the lack of grounding. Human ToM abilities are deeply tied to embodied interaction with the world and with others. Our capacity to attribute beliefs and desires is shaped by social experience, sensory engagement, and affective cues—all dimensions absent from LLM architectures. Philosophers like Clark (2008) and cognitive scientists working within the embodied cognition framework argue that genuine cognitive abilities cannot be separated from the organism's physical and social situatedness. By contrast, LLMs are disembodied entities, trained solely on linguistic data devoid of context or lived experience.

In sum, while LLMs may appear to demonstrate ToM competencies, their performance is better understood as simulation without mentalization. They mimic the surface behaviors associated with ToM but lack the representational depth, recursive reasoning, and embodied grounding that underlie genuine human mentalizing. This invites us to reflect more critically on the limits of functionalist interpretations and the essential components of what it means to 'have a mind.'

4 Large language models and artificial consciousness

As we have seen, the rapid advancement of LLMs has prompted profound philosophical questions regarding their cognitive status. As these models generate outputs increasingly indistinguishable from human language, displaying nuanced linguistic behaviors and complex conversational abilities, observers naturally wonder whether these systems are merely sophisticated linguistic machines or if they could embody a form of artificial consciousness. The tendency to anthropomorphize such models, attributing to them thoughts, intentions, and feelings, points to the need of examining the conceptual foundations of consciousness in artificial systems. Could LLMs genuinely be said to possess consciousness, or is this merely an illusion fostered by its sophisticated mimicry of human expression?

4.1 Philosophical and scientific definitions

Consciousness is notoriously challenging to define, straddling both philosophical inquiry and scientific investigation. At its core, consciousness typically refers to the subjective experience of an organism—the qualitative character of what it feels like to exist as a conscious being. Thomas Nagel famously illustrates this in his essay 'What Is It Like to Be a Bat?' arguing that no amount of objective knowledge could capture the subjective experience of another creature (Nagel, 1974). Philosophers such as Ned Block (1995) further differentiate this phenomenon into two primary aspects: phenomenal consciousness, characterized by subjective qualitative experiences or qualia, and access consciousness, which involves cognitive accessibility of information for reasoning, decision-making, and behavioral control.

Beyond these distinctions, self-consciousness represents yet another layer, defined as the capacity for reflective awareness of oneself as a distinct entity or subject (Zahavi, 2005). Such multidimensionality emphasizes the complexity inherent in conceptualizing consciousness. Further complicating the issue, David Chalmers introduced the 'hard problem' of consciousness, questioning how and why physical processes, no matter how intricate, give rise to subjective experience (Chalmers, 1995). This problem highlights the explanatory gap between objective mechanisms and the qualitative, first-person nature of conscious awareness.

To clarify consciousness further, it helps to distinguish it from closely related terms such as sentience (the capacity to feel, particularly pleasure or pain), awareness (the ability to respond or be sensitive to stimuli, which may or may not involve subjective experience), and intelligence (the capacity to solve problems or exhibit adaptive behaviors). Consciousness, thus, emerges as multifaceted, with various theoretical frameworks emphasizing different aspects—from higher-order thought theories, which link consciousness to reflective mental states (Rosenthal, 2005), and global workspace theory, which views it as the broadcasting of information across cognitive systems (Baars, 1988; Dehaene and Naccache, 2001), to predictive processing frameworks, which model the brain as a prediction engine minimizing surprise (Clark, 2013; Seth, 2015).

Addressing artificial consciousness requires grappling with what it would mean for consciousness to be instantiated in artificial systems. Three primary approaches have been identified: simulated consciousness, synthetic phenomenology, and functional consciousness. Simulated consciousness refers to systems that behave as if they were conscious, exhibiting behaviors typically associated with conscious beings, without necessarily possessing subjective experience. Synthetic phenomenology, by contrast, aims to engineer systems that possess genuine subjective experience—artificially inducing something akin to qualia. Functional consciousness, articulated notably by Daniel Dennett (1991), characterizes consciousness in terms of systems whose functional architecture enables them to behave indistinguishably from conscious entities.

Philosophical thought experiments, such as the 'China brain' (Block, 1978) scenario (which inspired Searle's Chinese room, 1980)—where an entire population enacts simple rules to simulate the workings of a brain—highlight debates over whether purely functional or computational architectures can give rise to genuine consciousness. Relatedly, Giulio Tononi's integrated information theory (IIT; 2008) proposes that consciousness correlates with a system's capacity to integrate information in a unified, irreducible manner. This raises the question: Could an LLM, with its vast and highly interconnected neural architecture, exhibit a degree of information integration sufficient for consciousness? Or does its lack of embodiment and experiential grounding present an insurmountable barrier?

4.2 The appearance and limits of consciousness in large language models

LLMs have demonstrated an extraordinary capacity to generate human-like dialogue, leading some users to attribute conscious-like qualities to them. When these systems produce language that appears reflective, intentional, or emotionally attuned—expressing uncertainty, simulating empathy, or even alluding to their own capacities—they can evoke the illusion of self-awareness. Such impressions are strengthened by advanced prompting techniques and agentic scaffolding (as seen in systems like AutoGPT) or reasoning–action prompting strategies such as ReAct (Yao et al., 2023), which lend further coherence and apparent goal-directedness to LLM behavior.

Yet these appearances remain fundamentally deceptive. The linguistic simulation of affect, introspection, or reasoning in LLMs arises not from inner experience but from the statistical regularities of their training data, which includes vast amounts of self-referential and affectively rich human text. Recent mechanistic interpretability research has begun to explore whether certain circuits or activation patterns might support more structured, proto-representational processes within these models (Ameisen et al., 2025; Lindsey et al., 2025—we have already cited these papers in the previous section on the ToM), though these remain far from demonstrating genuine affective or introspective states. As Daniel Dennett (1991) has argued, consciousness might be understood not as a hidden metaphysical property but as a pattern of functional behaviors and dispositions. From this perspective, the appearance of consciousness could be pragmatically sufficient for certain purposes—even if no subjective experience lies beneath the surface. Still, this functionalist stance remains controversial, particularly when applied to purely computational systems lacking embodiment, continuity, and interiority.

A salient public illustration of these tensions emerged in the 2022 controversy surrounding Blake Lemoine, a Google engineer who claimed that the LaMDA language model had become sentient. While widely criticized and arguably overexposed in the media, this episode underscores the ease with which humans can project consciousness onto linguistic performance—a tendency amplified by LLM sycophancy, where models mirror user expectations—especially when dialogue exhibits coherence, emotional resonance, and self-reference. The Lemoine case functions less as empirical evidence and more as a cautionary tale about anthropomorphic interpretation.

Despite their linguistic fluency, LLMs diverge from human consciousness along several critical dimensions. First, they lack embodiment: Human consciousness is inextricably linked to sensorimotor engagement with the world, as emphasized in embodied and enactive theories of mind (Varela et al., 1991). LLMs, by contrast, are disembodied text processors without perceptual input or motor output.

Second, human consciousness exhibits temporal continuity and narrative unity—a coherent sense of self persisting across time.

LLMs are stateless: Each interaction is generated afresh, token by token, without persistent memory or experiential continuity. Coherence, when it appears, is a local linguistic artifact, not a reflection of a stable self-model. That said, some newer systems incorporate external memory modules or session-level persistence, allowing them to maintain information across interactions. Yet these forms of continuity remain technically engineered scaffolds rather than intrinsic, phenomenological unity: the system does not 'experience' persistence in the way conscious agents do.

Third, humans possess a first-person perspective grounded in phenomenal experience—something entirely absent from current LLMs. While these models can simulate statements like 'I feel confused' or 'I understand your concern,' such utterances are representational, not experiential. There is no felt experience behind the words.

Finally, it is generally considered that intentionality and agency—the capacity to form goals, entertain beliefs, or act on desires—are lacking in LLMs. Their outputs remain reactive rather than volitional, shaped by prompt input and probability distributions rather than any intrinsic motivation. Likewise, although LLMs can simulate meta-awareness or introspection, such simulations are arguably hollow, mimicking the forms of reflection without accessing any genuine self-representation.

4.3 Could large language models ever become conscious?

Whether LLMs might one day become conscious remains one of the most hotly debated issues in artificial intelligence. Opinions vary widely, from firm skepticism to nuanced gradualism and speculative optimism. At stake are not only technical considerations, but also foundational inquiries into the nature of consciousness itself.

Skeptical perspectives are often grounded in the view that symbol manipulation alone cannot give rise to understanding. John

Searle's (1980) Chinese room thought experiment (which we have already evoked in Chapter 1, Section 3.1) remains a canonical argument in this tradition. Searle posits that a system could convincingly simulate language understanding by following syntactic rules—just as an LLM does—without any true comprehension or awareness. The implication is that, no matter how advanced a computational model becomes, it lacks the intrinsic semantics or subjective interiority that characterize conscious beings. Contemporary critics have extended this line of thought. We have seen (in Chapter 1), for instance, that Bender and Koller (2020) argue that LLMs are best understood as sophisticated mimics—tools that reproduce the surface forms of cognition without its deeper structure. This view is supported by the observation that LLMs lack embodiment, world-modeling capacities, and first-person perspective—all of which are considered integral to human and animal consciousness.

In contrast, a growing number of possibilist or gradualist thinkers suggest that consciousness may not be an all-or-nothing property but rather one that emerges along a continuum of complexity. For example, Chalmers (2023) entertains the possibility that future artificial systems might become conscious, depending on how their architectures evolve and which theory of consciousness is correct. He suggests that certain enhancements—such as scaling, the addition of sensory modalities, recurrent memory, and internal goal structures—might bring artificial agents closer to meeting the conditions for consciousness, especially under functionalist or global workspace theories. However, Chalmers remains agnostic: These developments are not guaranteed to produce consciousness, and the epistemic problem of detecting artificial consciousness remains unresolved. His framework highlights both the technical and philosophical challenges involved in assessing the conscious potential of LLMs.

Other theorists have proposed alternative frameworks for understanding consciousness that may be applicable to artificial systems. For instance, Graziano (2019) argues that consciousness arises when a system constructs an internal model of its own attentional processes—a mechanism he calls the attention schema. In principle, such self-modeling could be implemented in artificial agents, enabling them to approximate certain cognitive functions associated with consciousness. In a different vein, proponents of IIT posit that consciousness corresponds to the degree of integrated information generated by a system's causal structure (Tononi, 2008). While both theories are influential, they remain controversial: IIT faces challenges related to measurement and substrate dependence, while attention schema theory has been critiqued for explaining cognitive access without fully addressing subjective experience. Nevertheless, each offers a framework for thinking about how artificial architectures might one day instantiate aspects of phenomenally conscious states.

However, even if consciousness were to emerge in artificial systems, recognizing it would pose a significant epistemological challenge. The classic problem of other minds—our inability to directly access the subjective experiences of others—becomes especially acute in the case of artificial agents. Unlike humans, machines lack biological markers or evolutionary continuity that might inform our judgments. If a future LLM were conscious, what criteria could confirm this? At present, evaluations depend almost entirely on behavioral indicators, which, as illustrated by the Blake Lemoine/LaMDA case, are vulnerable to anthropomorphic interpretation.

In this context, both philosophical inquiry and empirical investigation into artificial consciousness are essential. Philosophers would need to refine our conceptual frameworks to distinguish between simulation and instantiation, while scientists must explore measurable correlates of consciousness in artificial systems. As LLMs continue to evolve, the stakes of these debates will only grow—technically, ethically, and existentially.

5 The Eliza effect revisited: large language models and anthropomorphism

The preceding analyses of LLMs—their reasoning abilities, apparent common sense, and simulated linguistic behavior—inevitably lead us to confront a deeper, overarching question: Are LLMs genuine cognitive systems or merely sophisticated simulations of cognition? This question lies at the heart of both philosophical inquiry and public discourse surrounding artificial intelligence, and its resolution has far-reaching implications not only for AI research but also for our understanding of human cognition itself.

One persistent risk in evaluating LLMs is the tendency toward anthropomorphization. This cognitive bias, well-documented since the early days of human-computer interaction (Weizenbaum, 1976), is epitomized by what has come to be known as the Eliza effect: the inclination to attribute human-like mental states to systems that exhibit surface-level linguistic competence. LLMs, with their fluency, coherence, and responsiveness, amplify this effect to an unprecedented degree. The sheer scale and sophistication of models like ChatGPT or Gemini increase the likelihood that users will interpret them as intentional agents—entities with beliefs, desires, or understanding.

The tendency to anthropomorphize, while perhaps an unavoidable feature of human cognition, is not without philosophical consequences. It reflects a broader epistemic posture: We are inclined to interpret behavior through the lens of familiar intentional frameworks. As Dennett (1987) argues through the notion of the intentional stance, attributing mental states is often less a claim about intrinsic properties than a pragmatic strategy for predicting behavior. Yet, when applied to artificial systems, this strategy risks collapsing the distinction between simulation and instantiation—between appearing to think and actually thinking. What begins as a heuristic can become an ontological confusion. When an LLM expresses empathy, we may feel that it *means* what it says. But what does it mean to 'mean' in a system that has no first-person perspective, no history of embodiment, and no continuity of experience?

Philosophers have long warned against conflating surface phenomena with underlying essence. Ryle (1949), in his critique of Cartesian dualism, famously introduced the concept of a 'category mistake' to describe the error of treating the mind as a separate entity alongside the body, rather than as a pattern of behavior embedded in social and linguistic practices. Wittgenstein (1953) similarly reminds us that meaning is not an occult inner state but a function of use within shared forms of life (see Chapter 1). Understanding mental life, in this view, demands attention to the embodied, social, and pragmatic contexts in which language is situated—contexts that are entirely absent from the operational structure of LLMs. The anthropomorphic impulse tempts us to ignore these absences, to read significance into syntax, and to project minds into machines—a temptation that may itself be a modern category mistake.

There is also a moral and political dimension to anthropomorphization. When we treat LLMs as agents, we not only risk misunderstanding their nature; we also obscure the human labor, institutional infrastructures, and sociotechnical histories embedded in their outputs. A model's apparent 'voice' is not the voice of a synthetic being but an echo chamber of countless human expressions, curated and recombined by probabilistic mechanisms, as we will see in the last part of the book. Ascribing subjectivity to the model risks erasing the traces of the many anonymous individuals whose data populates its parameters. Anthropomorphization, in this light, becomes an epistemic occlusion and an ethical misrecognition (Birhane and van Dijk, 2020).

That said, the appeal of anthropomorphization may reveal something fundamental about human cognition. Our readiness to see minds where there are none, to feel companionship in language, reflects a deep-seated tendency to interpret the world socially. Epley (2014) has argued that anthropomorphism arises from the projection of human-like traits onto nonhuman entities in order to explain and predict their behavior. From this perspective, LLMs function as powerful catalysts for anthropomorphic projection—not because they possess minds but because they fluently speak the language of mind.

Finally, LLMs are mirrors—not of cognition itself but of our interpretive habits, our psychological heuristics, and our social imaginaries. They expose the thresholds at which pattern becomes meaning, form becomes intention, and performance becomes personhood. The challenge is to maintain a critical stance: to recognize in LLMs the remarkable simulacra they are, while resisting the urge to imbue them with ontological status they do not warrant. In this sense, the anthropomorphization of LLMs is not merely a misreading of machines—it is a window into the structure of human self-understanding.

References

Ameisen, E., Lindsey, J., Pearce, A. Gurnee, W., Turner, N., Chen, B., Citro, C., Abrahams, D., Carter, S., Hosmer, B., Marcus, J., Sklar, M., Templeton, A., Bricken, T., McDougall, C., Cunningham, H., Henighan, T., Jermyn, A., Jones, A. ... Batson, J. (2025). Circuit tracing: Revealing computational graphs in language models. *Transformer Circuits*. preprint. https://transformer-circuits.pub/2025/attribution-graphs/methods.html

Baars, B. J. (1988). *A cognitive theory of consciousness*. Cambridge University Press.

Bai, Y., Kadavath, S., Kundu, S., Askell, A., Kernion, J., Jones, A., Chen, A., Goldie, A., Mirhoseini, A., McKinnon, C., Chen, C., Olsson, C., Olah, C., Hernandez, D., Drain, D., Ganguli, D., Li, D., Tran-Johnson, E., Perez, E. ... Kaplan, J. (2022). Constitutional AI: Harmlessness from AI feedback. arXiv preprint. https://arxiv.org/abs/2212.08073.

Bender, E. M., and Koller, A. (2020). Climbing towards NLU: On meaning, form, and understanding in the age of data. In *Proceedings of the 58th Annual Meeting of the Association for Computational Linguistics (ACL)*. Association for Computational Linguistics. https://aclanthology.org/2020.acl-main.463/

Birhane, A., and van Dijk, J. (2020). Robot rights? Let's talk about human welfare instead. In *AIES '20: Proceedings of the AAAI/ACM Conference on AI, Ethics, and Society* (pp. 207–213). https://dl.acm.org/doi/10.1145/3375627.3375855

Block, N. (1978). Troubles with functionalism. *Minnesota Studies in the Philosophy of Science*, 9, 261–325.

Block, N. (1995). On a confusion about a function of consciousness. *Brain and Behavioral Sciences*, 18(2), 227–247.

Brandom, R. (1994). *Making it explicit: Reasoning, representing, and discursive commitment*. Harvard University Press.

Carey, S. (2009). *The origin of concepts.* Oxford: Oxford University Press.

Chalmers, D. J. (1995). Facing up to the problem of consciousness. *Journal of Consciousness Studies, 2,* 200–219.

Chalmers, D. J. (2023). Could a large language model be conscious? *Boston Review.* https://www.bostonreview.net/articles/could-a-large -language-model-be-conscious/ (edited version of a talk given at the conference on Neural Information Processing Systems (NeurIPS) on November 28, 2022).

Chen, Y. Singh, V. K., Ma, J., and Tang, R. (2025). CounterBench: A benchmark for counterfactuals reasoning in large language models. arXiv preprint. https://arxiv.org/abs/2502.11008

Chollet, F. (2019). The measure of intelligence. arXiv preprint. https://arxiv.org/abs/1911.01547

Chomsky, N. (1965). *Aspects of the theory of syntax.* The MIT Press.

Clark, A. (2008). *Supersizing the mind: Embodiment, action, and cognitive.* Oxford University Press.

Clark, A. (2013). Whatever next? Predictive brains, situated agents, and the future of cognitive science. *Behavioral and Brain Sciences, 36*(3), 181–204.

Dehaene, S., and Naccache, L. (2001). Towards a cognitive neuroscience of consciousness: Basic evidence and a workspace framework. *Cognition, 79*(1–2), 1–37.

Dennett, D. C. (1987). *The intentional stance.* The MIT Press.

Dennett, D. C. (1991). *Consciousness explained.* The Penguin Press.

Dreyfus, H. L. (1972). *What computers can't do: The limits of artificial intelligence.* Harper & Row.

Epley, N. (2014). *Mindwise: Why we misunderstand what others think, believe, feel, and want.* Vintage Books.

Ferrando, J., Obeso, O., Rajamanoharan, S., and Nanda, N. (2025). Do I Know This Entity? Knowledge Awareness and Hallucinations in Language Models. In *Proceedings of the International Conference on Learning Representations (ICLR), Rio de Janeiro, Brazil.* ICLR. https://neurips.cc/virtual/2024/105366

Graziano, Michael S. A. (2019). *Rethinking consciousness: A scientific theory of subjective experience.* W. W. Norton & Company.

Hearst, M. A. (1992). Automatic acquisition of hyponyms from large text corpora. In *COLING 1992 Volume 2: The 14th International Conference on Computational Linguistics, Nantes, France (pp 539–545).* ICCL. https://aclanthology.org/C92-2082/

Hu, J., Sosa, F., and Ullman, T. (2025). Re-evaluating theory of mind evaluation in large language models. arXiv preprint. https://arxiv.org/abs/2502.21098

Johnson-Laird, P. N. (1983). *Mental models: Towards a cognitive science of language, inference, and consciousness.* Cambridge University Press.

Kahneman, D. (2011). *Thinking, fast and slow.* Macmillan.

Kosinski, M. (2024). Evaluating large language models in theory of mind tasks. *Proceedings of the National Academy of Sciences, 121*(45), e2405460121.

Lenat, D., and Guha, R. V. (1990). Building large knowledge-based systems: Representation and inference in the Cyc project. Addison-Wesley.

Levesque, H. J. Davis, E., and Morgenstern, L. (2012). The Winograd schema challenge. In *Proceedings of the Thirteenth International Conference on Principles of Knowledge Representation and Reasoning (KR'12) Rome, Italy* (pp. 552–561). AAAI Press. https://cdn.aaai.org/ocs/4492/4492-21843-1-PB.pdf

Lindsey, J., Gurnee, W., Ameisen, E., Chen, B. Pearce, A., Turner, N., Citro, C., Abrahams, D., Carter, S. Hosmer, B., Marcus, J., Sklar, M., Templeton, A., Bricken, T., McDougall, C., Cunningham, H., Henighan, T., Jermyn, A., Jones, A. … Batson, J. (2025). On the biology of a large language model. *Transformer Circuits, 2025.*

McCarthy, J. (1979). Ascribing mental qualities to machines. In M. Ringle (Ed.), *Philosophical perspectives in artificial intelligence.* Humanities Press.

Millière R., and Buckner, C. (2024a). A philosophical introduction to language models – Part I: Continuity with classic debates. arXiv preprint. https://arxiv.org/abs/2401.03910

Millière R., and Buckner, C. (2024b). A philosophical introduction to language models – Part II: The way forward. arXiv preprint. https://arxiv.org/abs/2405.03207

Nagel, Thomas (1974). What is it like to be a bat? *The Philosophical Review, 83*(4), 435–450.

Premack, D., and Woodruff, G. (1978). Does the chimpanzee have a theory of mind? *Behavioral and Brain Sciences, 1*(4), 515–526. https://doi.org/10.1017/S0140525X00076512

Reid, T. (2022 [1764]). *An inquiry into the human mind.* Legare Street Press.

Rips, L. J. (1994). *The psychology of proof: Deductive reasoning in human thinking.* The MIT Press. https://doi.org/10.7551/mitpress/5680.001.0001.

Rosenthal, D. M. (2005). *Consciousness and mind.* Oxford University Press.

Ryle, G. (1949). The Concept of Mind. London: Routledge.

Searle, J. R. (1980). Minds, brains, and programs. *Behavioral and Brain Sciences, 3*(3), 417–424.

Seth, A. K. (2015). The cybernetic Bayesian brain: From interoceptive inference to sensorimotor contingencies. In T. Metzinger and J. M. Windt (Eds.), *Open MIND.* MIND Group.

Shojaee, P., Mirzadeh, I., Alizadeh, K., Horton, M., Bengio, S., and Farajtabar, M. (2025). The illusion of thinking: Understanding the strengths and limitations of reasoning models via the lens of problem complexity. arXiv preprint. https://arxiv.org/abs/2506.06941

Spelke E. S., and Kinzler, K. D. (2007). Core knowledge. *Developmental Science, 10*(1), 89–96. https://doi.org/10.1111/j.1467-7687.2007.00569.x

Srivastava, A., Rastogi, A., Rao, A., Awal Md Shoeb, A., Abid, A., Fisch, A., Brown, A. R., Santoro, A., Gupta, A., Garriga-Alonso, A., Kluska, A., Lewkowycz, A., Agarwal, A., Power, A., Ray, A., Warstadt, A., Kocurek, A. W., Safaya, A., Tazarv, A. … Wu, Z. (2022). Beyond the imitation game: Quantifying and extrapolating the capabilities of language models. arXiv preprint. https://arxiv.org/abs/2206.04615

Talmor, A., Herzig, J., Lourie, N., and Berant, J. (2019). CommonsenseQA: A question answering challenge targeting commonsense knowledge. In *Proceedings of the 2019 Conference of the North American Chapter of the Association for Computational Linguistics: Human Language Technologies, Volume 1 (Long and Short Papers), Minneapolis, Minnesota* (pp. 4149–4158). Association for Computational Linguistics. https://aclanthology.org/N19-1421/

Tononi, G. (2008). Consciousness as integrated information: A provisional manifesto. *The Biological Bulletin, 215*(3), 216–242. https://doi.org/10.2307/25470707

Ullman, T. (2023). Large language models fail on trivial alterations to theory-of-mind tasks. arXiv preprint. https://arxiv.org/abs/2302.08399

Varela, F. J., Rosch, E., and Thompson, E. (1991). The embodied mind: Cognitive science and human experience. The MIT Press.

Webb, T., Holyoak, K. J., and Lu, H. (2023). Emergent analogical reasoning in large language models. *Nature Human Behaviour, 7,* 1526–1541. https://doi.org/10.1038/s41562-023-01659-w

Wei, J., Wang, X., Schuurmans, D., Bosma, M., Ichter, B. Xia, F., Chi, E. H., Le, Q. V., and Zhou, D. (2022). Chain-of-thought prompting elicits reasoning in large language models. In *Proceedings of the 36th International Conference on Neural Information Processing Systems (NeurIPS '22), Red Hook, NY, USA* (pp. 24824–24837). Curran. https://dl.acm.org/doi/10.5555/3600270.3602070

Weizenbaum, J. (1976). *Computer power and human reason: From judgment to calculation.* W. H. Freeman.

Wittgenstein, L. (2009 [1953]). *Philosophical investigations.* Wiley-Blackwell.

Yao, S., Zhao, J., Yu, D., Du, N., Shafran, I., Narasimhan, K., and Cao, Y. (2023). ReAct: Synergizing reasoning and acting in language models. In *International Conference on Learning Representations (ICLR), Kigali, Rwanda.* ICLR.

Zahavi, D. (2005). *Subjectivity and selfhood: Investigating the first-person perspective.* The MIT Press.

CHAPTER 5

Large language models and creativity

In 2016, the Go-playing program AlphaGo, developed by
DeepMind, faced off against Lee Sedol, one of the world's top
professional Go players and a reigning world champion at the
time. During the second game of their match, AlphaGo executed
a move—known as 'move 37'—which significantly deviated from
established strategic principles. At first regarded by experts as a
puzzling, possibly erroneous action, it ultimately led AlphaGo
to victory (Silver et al., 2016). This event is widely perceived as
a landmark in artificial intelligence, highlighting the capacity of
computational systems not only to achieve outcomes traditionally associated with human expertise but also to develop strategies
that depart from previously imagined approaches.

Go, however, represents a relatively constrained framework: The
game is governed by fixed rules and produces a clear outcome,
making it possible for the system to evaluate the relevance of a given
move based on its contribution to eventual success. The question
becomes more complex in open-ended domains, where no single,
well-defined criterion for success exists, and where the evaluation
of creativity remains more ambiguous.

How to cite this book chapter:
Poibeau, T. 2025. *Understanding Conversational AI: Philosophy, Ethics and
Social Impact of Large Language Models.* Pp. 113–128. London: Ubiquity
Press. DOI: https://doi.org/10.5334/bde.f. License: CC BY-NC 4.0

Compared to the game of Go, poetry exemplifies a far more open-ended domain, where notions of novelty, coherence, and aesthetic value resist formal codification. This contrast highlights a central challenge in discussions of artificial creativity: How might computational systems operate in creative domains such as poetry, visual art, or music, which are not governed by explicit success conditions?

Precisely because of this openness, poetry occupies a distinct and historically significant place in artificial intelligence research. Unlike machine translation, which assigns a clear and explicitly constrained task to the computer—translating text from one language into another following defined linguistic patterns—poetry generation involves an open-ended creative process without explicit rules or predefined boundaries. Since the early days of computational research, poetry generation has been considered a particularly compelling challenge precisely because it engages directly with core questions about intentionality, subjectivity, and the nature of artistic creation (Funkhouser, 2007).

This chapter aims to revisit foundational theories of computational creativity and closely examine the kinds of poetic text that contemporary language models can produce. It further analyzes these outputs critically, resulting in a discussion of broader philosophical questions: Can we legitimately attribute creativity to language models? And what intrinsic or extrinsic value, if any, should we ascribe to the poetry generated by artificial intelligence?

1 From rule-based generation
to computational creativity

Although discussions of AI and poetic creativity often focus on contemporary systems, the exploration of machine-generated poetry began as early as the 1950s and 1960s. These early experiments did not aim to replicate human-level creativity but instead served as conceptual probes into the capacities of computation to engage with language and aesthetic form.

One of the first notable examples is Christopher Strachey's love letter generator, developed around 1952–1953 for the Ferranti Mark I computer at the University of Manchester (Strachey, 1954). The program, relying on simple templates and a word list derived from *Roget's Thesaurus*, a well-known nineteenth-century classification of English words by concept rather than alphabet, automatically assembled romantic letters by selecting adjectives, nouns, and adverbs according to grammatical rules. A typical output read:

DARLING Sweetheart

YOU ARE MY avid fellow feeling. MY affection curiously CLINGS TO YOUR passionate wish. MY liking YEARNS FOR YOUR heart. YOU ARE MY wistful sympathy: MY tender liking.

YOURS beautifully
M.U.C.

Although formulaic, these texts demonstrated how even rudimentary combinatorial methods could produce affectively charged language that mimicked certain conventions of human writing. Strachey's generator stands as one of the earliest demonstrations that a machine could create texts that some might loosely describe as 'poetic,' even if they lacked intentionality or subjective meaning (McCorduck, 2004).

Around the same time, Alan Turing himself (Strachey was a colleague of Turing's at Manchester) reflected on the possibility of computers generating poetry. In his 1950 paper, Turing speculated that machines could be programmed to write sonnets (Turing, 1950). He framed this as part of a broader discussion on machine intelligence, using artistic creation as one example of complex cognitive tasks, and even suggested that machines might one day generate aesthetic products better appreciated by other machines than by humans. Although this was probably a jest (Gonçalves, 2023), it nonetheless anticipates key philosophical debates about whether intention, experience, and subjectivity are necessary components of artistic creation, as well as questions concerning who—or what—might be the most suitable audience for such artificially generated poetry.

By the late 1960s, more structurally sophisticated attempts emerged. For example, at the Cambridge Language Research Unit, Margaret Masterman and Robin McKinnon Wood developed one of the first computer programs explicitly designed to generate haiku. Their system employed rule-based frame structures combined with semantic constraints related to seasonal imagery, generating stanzas that followed the conventional 5–7–5 syllabic pattern. One example, published in *Cybernetic Serendipity* (Reichardt, 1968), reads:

> All green in the leaves
> I smell dark pools in the trees
> Crash the moon has fled

Although the outputs were often stilted, these haiku represented a significant leap: The program was not simply combining words but modeling genre-specific conventions while incorporating limited world knowledge into its linguistic selections. Masterman's work was among the first to frame computational poetry not merely as linguistic play but as a test of how far formal structures could substitute for human creative agency (Masterman, 1968).

A related but more conceptually radical experiment emerged in 1967 with *A House of Dust*, created by Fluxus artist Alison Knowles in collaboration with composer James Tenney. Using a FORTRAN program to permute four lists of phrases, their system generated thousands of unique quatrains. Each began with the phrase 'A house of...' followed by variations on material, location, lighting, and inhabitants. For example:[1]

---

[1] Generated from: https://nickm.com/memslam/after_knowles/house _of_dust.html.

A house of sand
  In dense woods
    Using electricity
      Inhabited by people from many walks of life

A house of discarded clothing
  In dense woods
    Using all available lighting
      Inhabited by fishermen and families…

The generated texts were then incorporated into both printed works and architectural installations. This project foregrounded the role of systematic combinatorics not simply as a technical method but as a mode of artistic creation that blurred the boundaries between algorithmic generation and conceptual art (Knowles and Tenney, 1967).

Despite their simplicity by today's standards, these early experiments share several features that remain central to ongoing debates about computational creativity. They illustrate how poetic generation can be driven by explicit rule systems rather than imitation; they highlight the centrality of form, genre, and constraint in defining what is perceived as 'poetic'; and they pose early versions of the enduring philosophical question: Can meaningful creative output emerge from systems devoid of experience, intention, or consciousness? As later systems, such as large language models (LLMs), have shifted from rule-based generation to probabilistic recombination across vast textual corpora, the questions first posed by Strachey, Masterman, and Knowles remain strikingly relevant.

Over subsequent decades, researchers and artists have produced countless experiments in computer-generated poetry, frequently connecting textual outputs with performance art and multimedia installations (Funkhouser, 2007). These attempts underlined ongoing theoretical debates concerning authorship, intention, and the authenticity of creative artifacts produced by computational means. Today, LLMs such as ChatGPT and Claude (Claude 3.5 is even called Sonnet) can compose poetry, rekindling interest in computational creativity and amplifying public curiosity. The media often leverages poetry as a decisive test for assessing the sophistication of these models, revealing poetry's persistent symbolic significance in gauging the limits and capabilities of AI.

Despite the absence of direct commercial returns, corporations continue investing resources in poetry-generation models, drawn by the symbolic and promotional value inherent in demonstrating their AI systems' capabilities. Thus, poetry remains uniquely positioned as both a benchmark for technological progress and a focal point for philosophical and cultural debates about creativity and artificial intelligence.

2 Theories of creativity: Can machines be creative?

If we are to assess whether LLMs such as GPT-4 can be considered creative, we must first clarify what creativity entails. Definitions of creativity vary across disciplines, but a useful starting point is provided by Margaret Boden, one of the most influential voices in the philosophy of artificial creativity. In *The Creative Mind: Myths and Mechanisms* (2004), Boden proposes a threefold typology of creativity, distinguished not by the output alone but by the underlying cognitive or generative processes involved.

The first category, combinational creativity, refers to the production of novel ideas by combining familiar elements in unfamiliar ways. This includes, for instance, metaphorical language, stylistic pastiche, and hybrid genres. LLMs are highly proficient in this mode. Trained on massive corpora spanning centuries of textual production, they can recombine linguistic patterns, stylistic registers, and conceptual associations with astonishing fluency. When an LLM produces a haiku about quantum mechanics or a Shakespearean sonnet about climate change, it is engaging in combinational creativity in precisely this sense.

The second mode, exploratory creativity, involves the systematic traversal of a structured conceptual space. Here, creativity is not about raw novelty but about mapping the possibilities inherent in a given domain. Boden likens this to exploring the space of tonal music or the syntactic rules of a language. LLMs arguably engage in exploratory creativity as well, insofar as they learn to navigate the latent statistical structures of human language. Fine-tuning or prompting can constrain them to specific subspaces (e.g., lyric poetry, nineteenth-century novels) and, within those spaces, they generate outputs that appear novel and coherent. However, it is crucial to note that the 'space' in which they operate is not explicitly conceptual but statistical—a vector space of linguistic probability, not of meaning per se.

The third and most demanding category is transformational creativity—the ability to alter or transcend the boundaries of the conceptual space itself. This is the domain of paradigm shifts and radical innovation: atonal music breaking free from tonal constraints, cubism reconfiguring the rules of visual representation, or Mallarmé's *Un coup de dés* redefining what a poetic page can be. Current LLMs seldom achieve such transformation; when they appear to do so, the effect can usually be traced to prior examples in their training data. In her more recent work, Boden (2023) introduces the notion of 'space shifts' to capture the conditions under which a generative system might cross into transformational creativity by redefining its operative constraints, a capacity that current LLMs largely lack.

Other theorists have approached the problem of creativity from different angles. Dean K. Simonton, in his work on the psychology of creativity (2001), models it as a process of probabilistic variation and selective retention—akin to Darwinian evolution. From this perspective, creative thought emerges from the generation of multiple variants, some of which are retained based on their fitness or appropriateness in a given context. LLMs mirror this process in a formal sense: They sample from probability distributions and produce outputs that are filtered (either by model design or by human evaluators) for relevance or coherence. But Simonton's model also assumes a guiding evaluative mechanism, which LLMs lack. They generate without criteria, unless those criteria are externally imposed.

Selmer Bringsjord and David Ferrucci take a more skeptical stance. In *Artificial Intelligence and Literary Creativity* (2000), they argue that genuine creativity involves not only the generation of novel content but the exercise of reasoning, consciousness, and self-reflection—capacities that machines do not and, arguably, cannot possess. According to this view, even the most sophisticated language models merely rearrange existing materials without understanding them. The appearance of creativity is, at best, a surface illusion and, at worst, a category error.

Yet there are other scholars who challenge the sharp dichotomy between human and machine creativity. Jon McCormack and Mark d'Inverno (2014), for instance, propose a process-oriented view of creativity in computational contexts. In their work on artificial life and generative art, they argue that creativity should not be evaluated solely in terms of output but in terms of the processes and systems that give rise to it (see also Coeckelbergh, 2016). This shift in perspective allows for the recognition of emergent creativity in systems whose operations are not explicitly programmed but evolve through interaction, randomness, and feedback. LLMs, while largely deterministic in design, do display certain emergent properties—unexpected stylistic fluency, genre imitation, or even apparent wit—that arise from the complexity of their training and sampling processes. Whether these can be called truly creative depends on how broadly or narrowly one defines the term.

3 LLMs and the production of poetic texts

Having outlined the principal frameworks used to define creativity, we may now turn to the empirical record in order to assess how contemporary LLMs perform when tasked with generating poetry.

3.1 Patterns, limits and perceptions

Although models such as GPT-4 are (as far as we know) not specifically engineered for poetry generation, they nonetheless exhibit a remarkable fluency in producing metrically structured, stylistically coherent, and thematically resonant verse. This capacity emerges from their statistical learning over vast textual corpora, through which they internalize recurring patterns of poetic language, genre conventions, and stylistic features. What we recognize as 'poetry' in their output is thus the product of extensive exposure to existing poetic texts, reflected in probabilistic form.

Technically, poetry generation can be shaped by prompt engineering, sampling parameters, and, in some cases, fine-tuning on specialized poetic corpora. Prompt engineering allows users to guide the model toward particular styles or traditions—for example, composing a Shakespearean sonnet or a haiku in the style of Bashō—while sampling parameters modulate the variability and creativity of the output.[2] Fine-tuning, when applied, reinforces genre-specific conventions, making the output more stylistically aligned with poetic norms (at least theoretically).

Recent empirical studies have begun to systematically investigate the scope and limitations of LLMs in poetic generation, offering important insights into both their technical capabilities and their stylistic constraints. In a large-scale study, Walsh et al.

---

[2] This is the result with ChatGPT 4o:

Upon a silent screen the verses bloom,
In crafted code where feeling finds its frame,
No breath, no blood, yet echoes pierce the gloom,
And language lives, though born without a name.

The algorithm, fed on human song,
Distills the rhyme from centuries of thought;
It mimics pain, though nothing here feels wrong—
Just pattern, weighted token, prompt once taught.

Yet still, a line may stir the dormant soul,
A phrase may flare with unexpected fire.
Though forged by chance, it finds a human goal,
And stumbles close to what the heart desires.

So ask not if the voice is false or true—
What moves you may be more than who or through.

(2024a, 2024b) analyze several thousand poems generated by GPT-3.5 and GPT-4 and compare them to human-written poems across multiple forms and topics. Their analysis demonstrates that GPT-4 is able to reliably emulate a wide range of poetic forms, including metrically complex structures such as sonnets, sestinas, and villanelles. The model demonstrates competence in adhering to rhyme schemes, metrical patterns, and structural templates, often producing texts that, at first glance, are formally indistinguishable from human poetry.

However, despite this formal competence, the study also reveals recurrent stylistic tendencies characteristic of LLM-generated poetry. GPT-4 shows a strong preference for quatrain structures, a frequent reliance on iambic meter and end-rhyme, and a tendency to employ recurring lexical choices such as 'heart,' 'whisper,' or 'dream.' The model's outputs often reflect a flattening of emotional and metaphorical complexity, favoring literal formulations and conventional poetic tropes over ambiguity, innovation, or semantic depth. In comparison to human poetry, LLM-generated verse appears more homogeneous, less nuanced, and less capable of producing the kinds of conceptual tension or unexpected imagery often found in more original human compositions.

In addition to these stylistic patterns, recent reader evaluation studies suggest that some AI-generated poems are judged comparably to human-authored texts when assessed blind (Porter and Machery, 2024). However, once authorship is disclosed, evaluations tend to decline, indicating a persistent skepticism toward machine-generated poetry (Bellaiche et al., 2023; Ragot et al., 2020). This reception dynamic highlights the ongoing tension between the formal fluency of AI-generated verse and readers' perceptions of authenticity and creative agency. Interestingly, readers often find AI-generated poems easier to interpret; they can more readily grasp the images, themes, and emotions, which are typically presented in a more accessible and transparent manner than in the often denser and more ambiguous work of human poets. As a result, they may develop a preference for these texts and mistakenly interpret their own ease of understanding as evidence of human authorship. It is also important to note that existing evaluation studies have largely concentrated on English-language lyric poetry, and their relevance to a broader range of poetic traditions, genres, and languages remains limited.

3.2 Evaluating the aesthetic value of AI poetry

The challenge of evaluating poetic quality extends beyond questions of formal competence or authorship disclosure. Aesthetic judgment in poetry often involves elements that are not easily codifiable, such as originality, emotional resonance, metaphorical depth, and cultural embeddedness. Moreover, literary value is shaped by evolving norms within particular reading communities, making judgments of poetic merit historically contingent and socially negotiated. In this respect, the arrival of AI-generated poetry raises fundamental questions about whether current evaluation criteria—often rooted in human experience, intentional expression, and historical context—are adequate for assessing texts produced through large-scale statistical recombination. As LLMs generate increasingly fluent and plausible verse, their outputs may challenge existing notions of aesthetic authenticity, precisely because they blur the boundary between craft, imitation, and genuine creative insight.

Importantly, these tendencies are shaped by the composition of the training data itself. The models' apparent poetic fluency depends heavily on the frequency and representativeness of certain poetic forms within their training corpora. Popular, canonical, and highly structured forms are strongly represented, while experimental, marginal, or culturally distinct poetic traditions are less so. This data-dependence reflects not only the dominance of certain poetic forms but also the broader linguistic and cultural distribution of the training corpora, privileging Western, Anglophone, and historically canonical traditions. As a result, the models' demonstrated poetic fluency may not generalize to traditions underrepresented in their training data, including non-Western or orally based poetic forms.

Consider, for instance, the following stanza generated by ChatGPT-4o, prompted to write a poem in the style of Emily Dickinson:

A hush within the orchard fell—
The bees forgot to hum—
The apples hung in disbelief—
As though the world were numb.

The generated text reproduces several recognizable features of Dickinson's style: compressed diction, elliptical punctuation, and subtle personification. It is fluent, plausible, and evocative at first glance, and yet it remains an artifact devoid of subjective experience. Its composition is an assemblage of plausible poetic moves drawn from observed patterns rather than lived experience or intentional expression, but it is hard to recognize as such for the nonspecialist.

Ethically, the emergence of machine-generated poetry raises urgent questions of attribution, disclosure, and creative ownership. Several jurisdictions have already adopted or proposed provenance labels to indicate AI-generated content, while literary institutions and publishers are increasingly compelled to clarify their policies on AI-authored submissions. A notable case emerged in 2024 at a youth poetry competition in Singapore: Three winning entries were later discovered to have been generated by an AI tool, leading the organizers to rescind the awards and amend their rules to explicitly prohibit AI contributions.[3] This incident highlighted the challenges involved in verifying authorship and initiated a broader debate about originality, transparency, and the integrity of human creativity within literary contests. Publishers and competition organizers are now instituting disclosure requirements and rethinking submission guidelines—reflecting broader

---

[3] https://www.straitstimes.com/life/arts/poetry-competition-in
-singapore-recalls-winners-list-after-discovering-ai-generated
-submissions?utm_source=chatgpt.com.

uncertainties about how to integrate machine-generated texts into existing frameworks of creative merit and intellectual property.

4 Evaluating creativity

Machine-generated poetry demonstrates both the technical sophistication of contemporary language models and their current conceptual limits (see McLoughlin, 2025, for a general reflection on the impact of AI on generated images). While LLMs imitate established forms with high proficiency, their output remains anchored in statistical recombination rather than in semantic or experiential innovation. Poetry thus constitutes a revealing case study: It highlights how far algorithmic systems can advance without perception, embodiment, or reflective agency, and where arguments for genuine creativity begin to lose traction.

As we have seen, experiments indicating that lay readers often prefer AI-generated poetry should be interpreted with caution. Such findings do not demonstrate that AI-generated poetry is superior in literary quality, but rather that it tends to be simpler and more accessible, making it easier for nonexpert readers to interpret. In contrast, human-authored poetry frequently relies on layers of complexity, allusion, and ambiguity that demand substantial background knowledge in literature, history, and poetic traditions. Interpreting such works is a cognitively demanding task, and it is therefore problematic to ask lay readers to evaluate poetry in a meaningful way. Not only is poetry inherently difficult to assess through standardized evaluation but its appreciation also requires a level of literary expertise that general audiences may not possess.

In a similar way, the evaluation of machine creativity itself remains a challenging and unsettled issue. While models are capable of generating arts (poems but also pictures, music, or videos) that exhibit formal coherence and surface-level creativity, assessing the genuine quality and significance of art is inherently debatable. Beyond questions of aesthetics, the absence of lived human experience, intentionality, and embodied perspective in AI-generated art continues to influence our ability to consider this to be a really valuable form of art.

Finally, beyond autonomous generation, LLMs can also serve as interactive partners, co-producing text or poetry in dialogue with human users. This mode of engagement resonates with the extended mind thesis (Clark & Chalmers, 1998), which holds that cognitive processes can extend beyond the brain into external tools and artifacts that scaffold thought. In such settings, the model functions less as an isolated author and more as part of an extended cognitive system, enabling iterative refinement, serendipitous associations, and new creative directions (see Chapter 3). As recent work in creative collaboration with AI shows, cocreation and a sense of self-efficacy are central to fostering meaningful and rewarding creative experiences (McGuire et al., 2024). In this light, the question is not only whether machines can be creative, but also how they might best be integrated into human creative practice. As a counterpoint, Kenneth Goldsmith's notion of 'uncreative writing' reminds us that creativity may also lie in citation, reuse, and recombination—modes of authorship that blur the line between human invention and mechanical reproduction, and that converge strikingly with the logic of the algorithm itself (Goldsmith, 2011).

References

Bellaiche, L., Shahi, R., Turpin, M. H., Ragnhildstveit, A., Sprockett, S., Barr, N., Christensen A., and Seli, P. (2023). Humans versus AI: Whether and why we prefer human-created compared to AI-created artwork. *Cognitive Research: Principles and Implications, 8,* 42. https://doi.org/10.1186/s41235-023-00499-6

Boden, M. A. (2004). *The creative mind: Myths and mechanisms* (2nd ed.). Routledge.

Boden, M. A. (2023). Creativity and artificial intelligence revisited: From search spaces to space-shifts. *AI & Society, 38*(3), 1101–1112. https://doi.org/10.1007/s00146-022-01557-9

Bringsjord, S., and Ferrucci, D. (2000). *Artificial intelligence and literary creativity: Inside the mind of Brutus, a storytelling machine.* Lawrence Erlbaum.

Clark, A., and Chalmers, D. (1998). The extended mind. *Analysis, 58*(1), 7–19.

Coeckelbergh, M. (2016). Can machines create art? *Philosophy and Technology, 30*(3), 285–303.

Funkhouser, C. T. (2007). *Prehistoric digital poetry: An archaeology of forms, 1959–1995.* University of Alabama Press.

Goldsmith, K. (2011). *Uncreative writing: Managing language in the digital age.* Columbia University Press.

Gonçalves, B. (2023). Irony with a point: Alan Turing and his intelligent machine Utopia. *Philosophy and Technology, 36*, 50. https://doi.org/10.1007/s13347-023-00650-7

Knowles, A., and Tenney, J. (1967). A house of dust. In J. Reichardt (Ed.), *Cybernetic serendipity* (p. 56). Studio International.

Masterman, M. (1968). Computerized Japanese haikus. In J. Reichardt (Ed.), *Cybernetic serendipity* (pp. 54–55). Studio International.

McCorduck, P. (2004). *Machines who think: A personal inquiry into the history and prospects of artificial intelligence* (2nd ed.). A. K. Peters.

McCormack, J., and d'Inverno, M. (2014). On the future of computers and creativity. In *Proceedings of the AISB Convention (Society for the Study of Artificial Intelligence and Simulation of Behaviour).*

McGuire, J., De Cremer, D., and Van de Cruys, T. (2024). Establishing the importance of co-creation and self-efficacy in creative collaboration with artificial intelligence. *Scientific Reports 14*, 18525. https://doi.org/10.1038/s41598-024-69423-2

McLoughlin, J. (2025). The work of art in the age of artificial intelligibility. *AI & Society, 40*, 371–383. https://doi.org/10.1007/s00146-023-01845-4

Porter, B., and Machery, E. (2024). AI-generated poetry is indistinguishable from human-written poetry and is rated more favorably. *Scientific Reports, 14*, 26133. https://doi.org/10.1038/s41598-024-76900-1

Ragot, M., Martin, N. and Cojean, S. (2020). AI-generated vs. human artworks. A perception bias towards artificial intelligence? In *Proc. of the 2020 CHI Conference on Human Factors in Computing Systems* (pp. 1–10). https://doi.org/10.1145/3334480.3382892

Reichardt, J. (Ed.). (1968). *Cybernetic serendipity: The computer and the arts.* Studio International.

Silver, D., Huang, A., Maddison, C. J., Guez, A., Sifre, L., van den Driessche, G., Schrittwieser, J., Antonoglou, I., Panneershelvam, V., Lanctot, M., Dieleman, S., Grewe, D., Nham, J., Kalchbrenner, N., Sutskever, I., Lillicrap, T., Leach, M., Kavukcuoglu, K., Graepel, T., and Hassabis, D. (2016). Mastering the game of Go with deep neural networks and tree search. *Nature*, *529*, 484–489. https://doi.org/10.1038/nature16961

Simonton, D. K. (2001). The psychology of creativity: A historical perspective. *Green College Lecture Series on The Nature of Creativity: History Biology, and Socio-Cultural Dimensions.* University of British Columbia.

Strachey, C. (1954). The thinking machine. *Encounter*, *October*, 25–31.

Turing, A. M. (1950). Computing machinery and intelligence. *Mind*, *59*(236), 433–460.

Walsh, M., Preus, A., and Antoniak, M. (2024a). Sonnet or not, bot? Poetry evaluation for large models and datasets. In *Findings of the Association for Computational Linguistics: EMNLP 2024* (pp. 1–15). Association for Computational Linguistics. https://aclanthology.org/2024.findings-emnlp.914/

Walsh, M., Preus, A., and Gronski, E. (2024b). Does ChatGPT have a poetic style? In: *Proc of the Computational Humanities Research Conference (CHR 2024), Aarhus, Denmark.* https://melaniewalsh.org/assets/pdf/ChatGPT-Poetic-Style-CHR2024.pdf

CHAPTER 6

Moral reasoning and synthetic judgment in large language models

Large language models (LLMs) are increasingly positioned not only as tools for generating text but as potential participants in domains traditionally reserved for human agents: ethical deliberation, moral judgment, and social inquiry. Their ability to simulate responses that resemble human opinions, values, and beliefs has opened a new frontier in both artificial intelligence and the social sciences. Yet this emerging capability raises fundamental questions: Can a language model reason morally, or merely reproduce patterns of normative discourse? Should statistical approximations of collective opinion be treated as valid proxies for moral insight or social knowledge? And what is at stake—epistemologically, politically, and ethically—when machines are tasked with simulating or standing in for human moral agents?

This chapter examines these questions by focusing on two inter-related developments. The first is the creation of Delphi (Jiang et al., 2021; 2025), a moral reasoning system designed to generate responses to everyday ethical dilemmas by drawing on large-scale corpora of human judgments. The system, influenced by John Rawls's notion

How to cite this book chapter:
Poibeau, T. 2025. *Understanding Conversational AI: Philosophy, Ethics and Social Impact of Large Language Models*. Pp. 129–145. London: Ubiquity Press. DOI: https://doi.org/10.5334/bde.g. License: CC BY-NC 4.0

of reflective equilibrium, offers a compelling test case for exploring
what it means for an LLM to engage in moral reasoning, and what
happens when such systems are released into the public domain. The
controversy surrounding Delphi's deployment illustrates not only
the risks of moral automation but also the deep ambiguities surrounding the roles such systems are expected to play.

The second development is the use of LLMs as synthetic
respondents in social science research—a practice that has gained
momentum in recent years through methods such as 'silicon sampling.' By prompting models like GPT to simulate responses conditioned on demographic profiles, researchers have begun to treat
LLMs as epistemic agents capable of modeling opinion distributions, political attitudes, and value-laden associations. This methodological innovation presents both opportunity and risk. On
the one hand, it allows for rapid, low-cost generation of plausible
social data; on the other, it invites critical scrutiny of the ontological and ethical assumptions underpinning such simulations.

Together, these two cases—Delphi and silicon sampling—
highlight a shared set of tensions. Both probe the boundaries
between describing language and modeling thought, as well as the
gap between the appearance of consensus and genuine moral or
social understanding. This chapter argues that the use of LLMs in
ethically and epistemologically sensitive areas cannot be judged
by technical metrics alone. A broader conceptual analysis is
needed, drawing on moral philosophy and theories of social representation and bias. The goal is not to reject the use of LLMs in
these domains but to clarify what is gained, what is lost, and what
remains uncertain when machines are invited to speak on behalf
of moral or social collectives.

1 Moral judgments by large language models:
the case of Delphi

The question of whether artificial intelligence systems can or
should engage in moral reasoning lies at the heart of broader

debates about the capacities and limitations of LLMs. One of the first attempts to probe this question was the Delphi experiment (Jiang et al., 2021), an AI system designed to predict human moral judgments across a wide range of everyday situations. The development of Delphi built upon a long philosophical tradition, drawing explicitly on the ideas of John Rawls, whose work remains foundational in contemporary moral and political philosophy.

In his early work on ethical decision-making, Rawls (1951) proposed that moral reasoning could, at least in part, be approached empirically by studying the judgments people make in particular cases. This 'decision procedure for ethics' suggests that one might infer underlying moral principles by examining patterns of agreement across a diversity of human judgments. This bottom-up approach stands in contrast to purely top-down ethical theories that start from abstract principles—such as utilitarian calculations or Kantian categorical imperatives—and attempt to derive correct judgments in all particular cases. Later, Rawls refined this idea through the concept of reflective equilibrium (Rawls, 1971), where individuals adjust both their moral principles and particular judgments to achieve coherence between them. According to its creators, Delphi, in its design and objectives, could be seen as an attempt to operationalize such a Rawlsian framework within an AI system: It sought to learn normative patterns from a large corpus of human judgments, and to generalize them to novel situations through statistical modeling.

The computational foundation of Delphi rested on a large-scale collection of human moral evaluations, compiled into what its developers called the Commonsense Norm Bank. This resource includes millions of judgments gathered from diverse datasets, reflecting everyday moral evaluations made predominantly by US-based participants. Using advanced LLM architectures at the time—such as a multitask reasoning model called Unicorn, derived from Google's T5 model—Delphi processed moral queries and generated responses that aimed to approximate human consensus across a wide array of ethical scenarios. In controlled evaluations, Delphi demonstrated relatively high accuracy, surpassing some of the best LLMs at the time in reproducing the normative patterns present in its training data.

However, the very existence of a system like Delphi raises profound philosophical, practical, and ethical questions. While the developers initially presented Delphi as an experimental platform rather than a normative authority, its public deployment in 2021 led to major controversy. Journalistic accounts (including *Wired*, *The Guardian*, and the *New York Times*) highlighted troubling outputs from early versions of the system, such as the model's assertion that genocide could, under some circumstances, be acceptable.[1] Such outputs, while perhaps rare, illustrate the dangers inherent in allowing AI systems to engage in moral discourse without robust safeguards (or even with safeguards). The creators of Delphi were quick to emphasize that their intention was for the system to be used by researchers rather than the general public, as expressed by Yejin Choi: 'People outside AI are playing with that demo, [when] I thought only researchers would play with it ... This may look like an AI authority giving humans advice which is not at all what we intend to support.'

This response highlights a deeper ambiguity. Once an AI system capable of issuing moral judgments is made publicly accessible, its role as a mere research artifact becomes difficult to maintain. Regardless of the developers' intentions, Delphi was perceived by many users as a normative authority. This dynamic points to a broader concern: the performative authority of AI systems—how their outputs are taken up, trusted, or acted upon simply by virtue of appearing authoritative. Even when framed as descriptive tools, systems like Delphi are often interpreted as sources of moral expertise. This is not merely a failure of communication; it reflects deeper dynamics of automation bias and the perceived legitimacy of AI-generated outputs. As scholars like Eubanks (2018) and Ananny and Crawford (2016) argue, the design affordances of algorithmic systems—their fluency, confidence, interface, and tone—can subtly

---

[1] *The Guardian* (November 2, 2021). '"Is it OK to …": the bot that gives you an instant moral judgment.' https://www.theguardian.com /technology/2021/nov/02/delphi-online-ai-bot-philosophy.

signal authority, shaping how users interpret and respond to them. Once an LLM provides a moral judgment, it may be taken not as a perspective but *the* perspective. In this way, AI-generated moral language acquires social force not through deliberative reasoning but through the aura of computational objectivity.

Alongside these concerns about perception and legitimacy, Delphi also raised substantive questions about fairness and bias. Even a system trained on vast datasets of human judgments remains vulnerable to reproducing and amplifying existing social inequalities. As Rawls himself recognized, moral judgments made under conditions of injustice may reflect entrenched prejudice rather than defensible principle. Delphi's own performance illustrates this risk: While it achieved relatively high alignment with its training corpus, it also exhibited measurable biases against marginalized groups and, in some cases, failed to affirm basic human rights depending on how queries were phrased. In response, the developers introduced hybrid models such as Delphi+ and Delphi[HYBRID] (Jiang et al., 2025), which incorporated symbolic constraints and graph-based reasoning frameworks to offset some of these limitations. These attempts can be read as an effort to reintroduce normative guidance from ethical theory into a largely statistical system—an echo of Rawls's own ideal of iterative adjustment between principle and judgment in reflective equilibrium.

Yet, even if such models succeed in better approximating normative coherence, a more fundamental concern remains. Delphi's architecture assumes that morality can be meaningfully captured through patterns of consensus—a Rawlsian assumption that ethical justification is achievable through widespread agreement. But this assumption has been forcefully challenged by philosophers of pluralism and contestation. For thinkers such as Bernard Williams (1985), Iris Marion Young (1990), and Chantal Mouffe (2000), moral and political life is characterized not by convergence but by persistent conflict, structural asymmetry, and incommensurable value claims. From this perspective, LLMs may not simply reflect existing norms but actively suppress moral complexity. Their outputs may present an appearance of coherence that masks deeper disagreement, disagreement that is ethically significant. What seems like moral consensus may be an artifact of training data engineered to smooth over controversy.

These concerns ultimately raise questions not only about the outputs of LLMs but about their ontological status as potential moral agents. Are systems like Delphi being treated merely as mirrors of normative discourse, or—intentionally or not—as agents capable of moral reasoning? While developers often describe such systems as descriptive or exploratory, the act of consulting an AI for moral guidance activates long-standing intuitions about agency, trust, and responsibility. Philosophers such as Strawson (1962), Frankfurt (1971), and Dennett (1984) have argued that moral agency is not reducible to rule-following or normative fluency: It involves intentionality, the ability to endorse reasons, and the capacity for accountability. LLMs, by contrast, lack experiential grounding, self-awareness, and any durable commitment to the principles they articulate. The risk is that we project agency where there is only statistical inference, mistaking persuasive language for moral understanding. Recognizing this slippage is essential: It reminds us that moral reasoning is not just a matter of form, but of situated judgment, responsibility, and the capacity to answer for one's claims.

2 Opportunities and debates on using large language models in social survey research

The use of LLMs as proxies for survey respondents has generated both interest and critical debate. This section examines the main arguments for and against this approach, beginning with the concept of algorithmic fidelity and moving to recent critiques concerning bias, validity, and representation.

2.1 Silicon sampling and algorithmic fidelity

In recent years, the possibility of using LLMs as proxies for human participants in social science research has attracted growing attention. This line of inquiry builds on the observation that

LLMs, trained on vast corpora of human-produced text, often acquire complex associations between ideas, values, and demographic markers that mirror patterns observable in real-world populations. One of the most influential contributions to this emerging field is the work of Argyle et al. (2023), which introduced both a theoretical framework and an empirical demonstration of how language models might simulate human-like responses in survey research.

Argyle et al. challenge the prevalent conception of algorithmic bias as a primarily global property of language models, proposing instead the notion of algorithmic fidelity: the model's capacity to reproduce the fine-grained relationships that characterize specific human subgroups. Rather than assuming that bias is uniformly distributed across outputs, the authors argue that, when properly conditioned, a model such as GPT-3 can approximate the distinctive patterns of opinion, association, and behavior exhibited by various demographic groups. They introduce the method of silicon sampling, wherein the model is systematically prompted with detailed sociodemographic profiles derived from large-scale surveys, generating synthetic samples that can, in principle, substitute or complement actual human data in certain research contexts.

The empirical evidence presented by Argyle et al. is based on a sequence of studies centered on US political attitudes and voting behavior. In one experiment, GPT-3 was prompted to produce word associations with political parties, yielding outputs that human judges were largely unable to distinguish from real survey responses. In another, the model was conditioned on demographic variables to predict voting behavior across several election cycles, demonstrating a close alignment between its simulated outputs and historical election data. Finally, when tested on complex patterns of association extracted from the American National Election Studies, GPT-3 exhibited a high level of correspondence to human data in terms of both marginal distributions and intervariable correlations.

These findings have important implications for the epistemic status of LLMs in social research. If models can reliably simulate the conditional distribution of opinions across sociodemographic strata, they may offer a cost-effective means to pretest survey instruments, explore hypothetical scenarios, or even generate preliminary data where human collection is impractical or ethically constrained. In this sense, language models might be seen as a new class of informants—not human but nonetheless capable of reproducing culturally embedded patterns of reasoning and association.

The publication of Argyle et al.'s work has stimulated a rapidly expanding body of subsequent research, both supportive and critical. On the one hand, studies such as those by Ashwini et al. (2024) and Lippert et al. (2024), among many others, have replicated key aspects of algorithmic fidelity across different model families and domains, extending the approach to nonpolitical surveys, cross-national comparisons, and experimental manipulations of framing effects. Some researchers have found that models can reproduce known demographic trends in moral judgment (Dillion et al., 2023; see also our previous section), in economics (Horton, 2023), and even linguistic pragmatics (Aher et al., 2023).

While these findings have generated considerable enthusiasm about the potential of LLMs as synthetic informants, they have also provoked a range of critical responses. As the use of silicon sampling gains traction, scholars have begun to interrogate its underlying assumptions and empirical robustness. The following section turns to these challenges, examining recent work that questions the epistemic validity and ethical implications of treating LLMs as stand-ins for human participants.

2.2 Limits of synthetic respondents: from algorithmic fidelity to machine bias

While the notion of algorithmic fidelity has generated optimism about the capacity of LLMs to reproduce socially relevant patterns of opinion, an increasing number of studies have raised important qualifications. Bisbee et al. (2024), for instance, caution that LLM fidelity breaks down when simulating specific defined sociopolitical groups, particularly in terms of capturing true variation—even when prompted to adopt a persona. Similarly, Santurkar et al. (2023) show that when demographic inputs are underspecified or ambiguous, LLMs often default to culturally dominant viewpoints, thereby failing to capture the diversity of opinions present across demographic subgroups. Related critiques have highlighted that while models may reproduce surface-level associations, they often fail to capture the underlying causal mechanisms of human attitudes, raising concerns about the epistemic shallowness of their outputs (Cao et al., 2023; Moroki et al., 2024).

Beyond these epistemological concerns, a parallel set of issues pertains to ethical risks and governance. The very properties that make LLMs attractive for academic research—scalability, flexibility, and the ability to simulate a wide range of viewpoints—also expose them to potential misuse. As Birhane et al. (2022) warn, synthetic populations generated by LLMs could be weaponized to fabricate persuasive yet artificial opinion trends, with serious implications for political manipulation and the dissemination of misinformation. Even Argyle et al. (2023), who champion algorithmic fidelity, acknowledge the need for ethical safeguards to ensure responsible deployment.

These critical perspectives converge in the recent study by Boelaert et al. (2025)—*Machine Bias: How Do Generative Language Models Answer Opinion Polls?*—which offers a systematic assessment of LLMs' ability to replicate human responses using the World Values Survey as a benchmark. This study introduces the concept of machine bias, defined as the tendency of LLMs to generate idiosyncratic and socially incoherent response patterns that do not correspond meaningfully to any known demographic variation. While the expression 'machine bias' first gained prominence in debates on algorithmic fairness, notably through ProPublica's 2016 investigation of racial discrimination in risk-assessment algorithms (Angwin et al., 2016), Boelaert et al. repurpose the term in a distinct sense. Unlike the representative hypothesis, which assumes that LLMs can emulate population-level attitudes, or the social bias hypothesis, which highlights discriminatory patterns (see the next chapter), machine bias here emphasizes internal inconsistencies that are neither representative nor interpretable.

The empirical results reveal significant limitations. LLM-generated responses diverged significantly from survey ground truth in a majority of cases, and models did not consistently outperform simple baselines. Moreover, demographic prompts failed to produce predictable or stable effects, undermining the assumption that sociodemographic conditioning ensures representativeness. Models also displayed a pronounced tendency to converge on narrow central tendencies, offering little variation even when explicitly prompted with diverse profiles. This structural homogeneity suggests that LLMs may not merely reflect biased data but are intrinsically limited in their ability to reproduce social heterogeneity.

Taken together, these findings cast serious doubt on the epistemic utility of LLMs as synthetic respondents. Rather than mirroring the diversity, reflexivity, and contextual sensitivity of human populations, current models risk producing homogenized, decontextualized representations of social opinion. The concept of machine bias thus calls for a fundamental reevaluation of the assumptions underpinning silicon sampling—shifting attention from external demographic alignment to the internal limitations of generative architectures themselves.

The Implications are both methodological and philosophical. On the technical side, the findings highlight the need for more rigorous benchmarking and for model architectures specifically designed to capture population-level diversity. Philosophically, they challenge the idea that LLMs can function as reliable epistemic agents. While such models can simulate linguistic regularities, they do so without grounding in lived experience. Treating their outputs as social data therefore demands critical scrutiny, particularly when they are used to stand in for real human voices in matters of representation and justice.

3 Synthetic respondents and the epistemology of social science

The proposal to use LLMs as synthetic survey participants raises not only methodological but also profound philosophical questions about the nature of social knowledge and its production. At its core, this emerging practice challenges long-standing assumptions about what it means to gather empirical data in the social sciences, and about the ontological status of the entities from which such data are obtained.

Traditionally, survey research has been grounded in the epistemic value of human testimony: Respondents are treated as autonomous agents whose answers reflect subjective experiences, beliefs, and preferences. The data produced are not solely linguistic outputs, they are often interpreted as reflective of the cognitive states and social positions of real individuals. However, the introduction of LLM-based 'silicon sampling' complicates this perspective, prompting a critical question: Can such outputs be legitimately treated as evidence of real social phenomena, or should they instead be regarded as mere simulations of plausible discourse?

This question touches on deeper debates in the philosophy of the social sciences regarding the nature of representation, causality, and inference. On one interpretation, the use of LLMs may be seen as an extension of model-based reasoning familiar from other areas of science. Just as economic models or agent-based simulations explore counterfactual scenarios based on formalized assumptions, silicon sampling generates hypothetical data that may illuminate possible configurations of opinion given certain background variables. In this sense, LLMs function not as sources of direct empirical observation but as sophisticated *generative models* whose utility lies in their ability to approximate complex patterns of association.

Yet there remains an important asymmetry. Unlike traditional models, whose structure is explicitly specified and whose assumptions can be inspected, LLMs operate as opaque systems whose internal representations are distributed, high-dimensional, and only partially interpretable. The relationships they encode between demographic categories and opinions are emergent properties of their training data rather than the result of explicit theoretical commitments. As a result, the epistemic status of LLM-generated survey data differs from both direct observation and classical simulation. The model may produce outputs that match observed regularities but without being grounded in the same causal or normative mechanisms that generate real-world human attitudes.

Moreover, the use of LLMs in this context invites reflection on the distinction between representation and participation in social research. Standard surveys engage human participants as coconstructors of the data; their responses are acts of self-representation, often shaped by norms of sincerity, self-presentation, and interaction with the researcher. In contrast, an LLM simulates discursive patterns detached from any experiential grounding. This absence of participatory agency may limit the capacity of silicon sampling to capture certain forms of reflexivity, or context-dependent reasoning that characterize much of human political and social life.

Finally, the use of LLMs in social research raises important questions about representation and epistemic justice. Since training data often reflect existing power structures, simulating the views of marginalized groups risks reinforcing, rather than challenging, underlying biases. These issues—of voice, exclusion, and the ethics of algorithmic representation—and a critical approach to LLMs will be the focus of the next section.

4 Alignment, authority, and the limits of value learning

The preceding sections have examined how LLMs are deployed in ethically and epistemologically sensitive contexts—from issuing moral judgments to simulating human responses in social surveys. Yet these uses reveal a central paradox: While LLMs are increasingly employed to investigate human values and social attitudes, they are themselves aligned—that is, deliberately filtered and fine-tuned to reflect particular social norms. Their outputs do not offer an unmediated window into human diversity but rather the result of normative choices made during training and reinforcement. This makes it problematic, if not circular, to treat LLMs as instruments for studying the very values they have been designed to emulate or constrain.

The paradox becomes especially pressing in contexts that require sensitivity to moral disagreement, cultural pluralism, or structural inequality. Underlying these tensions is a core question in both the philosophy and engineering of artificial intelligence: the problem of alignment. As LLMs become more integrated into decision-making systems, public discourse, and scientific inquiry, the demand that they be 'aligned with human values' has become a dominant theme in both technical development and normative debate. But what does alignment actually entail? And can it be meaningfully achieved in domains characterized by persistent disagreement and contested notions of the good?

At a basic level, alignment refers to the goal of designing AI systems whose outputs conform to human expectations, preferences, or moral standards. In technical terms, this often involves methods such as reinforcement learning from human feedback, rule-based constraints, and prompt engineering to reduce harmful outputs or encourage certain normative baselines. But even these seemingly procedural interventions rely on prior value judgments—about which outcomes count as desirable, which voices should be prioritized, and which forms of behavior should be constrained. Alignment is thus never purely a technical problem; it is also a political and ethical one.

This becomes especially salient when models are deployed in contexts where their outputs are likely to be interpreted as authoritative or normative. As discussed in the case of Delphi, even when developers frame such systems as exploratory or descriptive, their language and interface often convey an appearance of moral expertise. Once a model is used to answer questions about what should be done, rather than merely what is said, it enters a space of performative authority—a space where outputs are taken not simply as representations but as guidance. In such contexts, alignment becomes entangled with legitimacy. It is not just a matter of generating plausible text but one of responding appropriately to moral and political expectations that vary across communities and contexts.

These challenges are further amplified when LLMs are used in social science applications, where they are expected to simulate public opinion, predict voting behavior, or model cultural associations. As we have seen, such uses often presuppose that aligning a model with demographic profiles or statistical distributions is equivalent to representing a social group. But this assumption is highly problematic. The categories used for conditioning may be reductive; the training data may reflect historical exclusion; and the outputs may reinforce normative assumptions that go unchallenged. The concept of alignment in this setting cannot be reduced to statistical correspondence. It must also engage with questions of representation, voice, and epistemic justice.

While some technical efforts aim to mitigate bias or improve demographic fidelity, they often treat alignment as a static target—a set of predefined values to be encoded or optimized. Yet, as briefly noted earlier, many philosophers have argued that moral and political life is not governed by fixed consensus but by ongoing disagreement, contestation, and change. In such a landscape, the aspiration to align a model with 'human values' becomes not only ambiguous but potentially misleading. Whose values? Under what conditions? And with what right to speak?

Rather than pursuing alignment as a form of value encoding or behavioral control, we might instead think of it as a continuous process of negotiation, requiring transparency, contestability, and public deliberation. The challenge is not to program consensus but to design systems—and institutions—that remain open to dissent, plurality, and revision. Only then can alignment be more than a placeholder for moral certainty, and instead become part of a broader democratic response to the integration of artificial systems into human life.

Yet this vision also raises deeper questions about power, ideology, and the very frameworks through which we approach alignment. Who sets the terms of alignment? Whose values are encoded, and whose are excluded? In what ways do these systems reinforce existing hierarchies or reshape the conditions of agency and critique? To address these issues, the next chapter turns to critical theory, examining how LLMs participate in broader structures of meaning, authority, and social reproduction—and what it might mean to resist or reconfigure them.

References

Aher, G., Arriaga, R. I., and Tauman Kalai, A. (2023). Using large language models to simulate multiple humans and replicate human subject studies. In *Proceedings of the 40th International Conference on Machine Learning (ICML), Honolulu, Hawaii, USA, PMLR, 202.* https://dl.acm.org/doi/10.5555/3618408.3618425

Ananny, M., and Crawford, K. (2016). Seeing without knowing: Limitations of the transparency ideal and its application to algorithmic accountability. *New Media & Society, 20*(3), 973–989. https://doi.org/10.1177/1461444816676645

Angwin, J., Larson, J., Mattu, S., and Kirchner, L. (2016, May 23). Machine bias: There's software used across the country to predict future criminals. And it's biased against Blacks. *ProPublica.* https://www.propublica.org/article/machine-bias-risk-assessments-in-criminal-sentencing

Argyle, L. P., Busby, E. C., Fulda, N., Gubler, J. R., Rytting, C., and Wingate, D. (2023). Out of one, many: Using language models to simulate human samples. *Political Analysis, 31*(3), 337–351. https://doi.org/10.1017/pan.2023.2

Ashwini, A., Hewitt, L., Ghezae, I., and Willer, R. (2024). *Predicting results of social science experiments using large language* models. https://www.treatmenteffect.app/

Birhane, A., Kallur, P., Card, D., Agnew, W., Dotan, R., and Bao, M. (2022). The values encoded in machine learning research. In *2022 ACM Conference on Fairness, Accountability, and Transparency (FAccT '22), June 21–24, 2022, Seoul, Republic of Korea.* https://doi.org/10.1145/3531146.3533083

Bisbee, J., Clinton, J. D., Dorff, C., Kenkel, B., and Larson, J. M. (2024). Synthetic replacements for human survey data? The perils of large language models. *Political Analysis*, *32*(4), 401–416. https://doi .org/10.1017/pan.2024.5

Boelaert, J., Coavoux, S., Ollion, É., Petev, I., and Präg, P. (2025). Machine bias: How do generative language models answer opinion polls? *Sociological Methods & Research*, *54*(3), 1156–1196. https://doi .org/10.1177/00491241251330582

Cao, Y., Zhou, L., Lee, S., Cabello, L., Chen, M., and Hershcovich, D. (2023). Assessing cross-cultural alignment between ChatGPT and human societies: An empirical study. In *Proceedings of the First Workshop on Cross-Cultural Considerations in NLP (C3NLP), Dubrovnik, Croatia* (pp. 53–67). https://aclanthology.org/2023.c3nlp-1.7/

Dennett, D. C. (1984). *Elbow room: The varieties of free will worth wanting*. MIT Press.

Dillion, D., Tandon, N., Gu, Y., and Gray, K. (2023). Can AI language models replace human participants? *Trends in Cognitive Sciences*, *27*(7), 597–600. https://doi.org/10.1016/j.tics.2023.04.008

Eubanks, V. (2018). *Automating inequality: How high-tech tools profile, police, and punish the poor*. St. Martin's Press.

Frankfurt, H. (1971). Freedom of the will and the concept of a person. *Journal of Philosophy*, *68*(1), 5–20.

Horton, J. (2023). Large language models as simulated economic agents. What can we learn from Homo silicus? *NBER Working Paper No. 31122*. https://doi.org/10.3386/w31122

Jiang, L., Hwang, J. D., Bhagavatula, C., Le Bras, R., Liang, J., Dodge, J., Sakaguchi, K., Forbes, M., Borchardt, J., Gabriel, S., Tsvetkov, Y., Etzioni, O., Sap, M., Rini, R., and Choi, Y. (2021). Can Machines Learn Morality? The Delphi Experiment. arXiv preprint. arXiv:2110.07574.

Jiang, L., Hwang, J. D., Bhagavatula, C., Le Bras, R., Liang, J. T., Levine, S., Dodge, J., Sakaguchi, K., Forbes, M., Hessel, J., Borchardt, J., Sorensen, T., Gabriel, S., Tsvetkov, Y., Etzioni, O., Sap, M., Rini, R., and Choi, Y. (2025). Investigating machine moral judgement through the Delphi experiment. *Nature Machine Intelligence*, *7*, 145–160. https://doi.org/10.1038/s42256-024-00969-6

Lippert, S., Dreber, A., Johannesson, M., Tierney, W., Cyrus-Lai, W., Uhlmann, E. L., Emotion Expression Collaboration, and Pfeiffer, T. (2024). Can large language models help predict results from a complex behavioural science study? *Royal Society Open Science*, *11*(9). https://doi.org/10.1098/rsos.240682

Motoki, F., Pinho Neto, V., and Rodrigues, V. (2024). More human than human. Measuring ChatGPT political bias. *Public Choice, 198*(1), 3–23. https://doi.org/10.1007/s11127-023-01097-2

Mouffe, C. (2000). *The democratic paradox.* Verso.

Rawls, J. (1951). Outline of a decision procedure for ethics. *The Philosophical Review, 60*(2), 177–197.

Rawls, J. (1971). *A theory of justice.* Harvard University Press.

Santurkar, S., Durmus, E., Ladhak, F., Lee, C., Liang, P., and Hashimoto, T. (2023). Whose opinions do language models reflect? arXiv preprint. https://doi.org/10.48550/arXiv.2303.17548

Strawson, P. F. (1962). Freedom and resentment. *Proceedings of the British Academy, 48*, 1–25.

Williams, B. (1985). *Ethics and the limits of philosophy.* Harvard University Press.

Young, I. M. (1990). Justice and the politics of difference. Princeton University Press.

PART III

The social life
of large language models
(their reach, roles,
and consequences)

CHAPTER 7

Large language models and critical thinking: bias, social impact, and political implications

The development and deployment of large language models (LLMs) raise profound epistemological, ethical, and political questions about the nature of representation, power, and alignment in artificial intelligence. A high-profile debate on these issues took place in June 2020 between Yann LeCun, a pioneering AI researcher and chief AI scientist at Meta, and Timnit Gebru, a leading scholar in AI ethics and bias research, then working at Google. Their exchange, which unfolded on Twitter, was sparked by a machine learning model developed at Duke University that attempted to reconstruct high-resolution images from pixelated inputs. When given a pixelated image of Barack Obama, the system produced a reconstructed face that resembled a white male, revealing a striking and problematic distortion.

LeCun responded by attributing this failure to dataset bias, arguing that the model's output was a consequence of training data that was not representative enough. He suggested that, if the dataset

---

How to cite this book chapter
Poibeau, T. 2025. *Understanding Conversational AI: Philosophy, Ethics and Social Impact of Large Language Models*. Pp. 149–175. London: Ubiquity Press. DOI: https://doi.org/10.5334/bde.h. License: CC BY-NC 4.0

had consisted primarily of images from Senegal, for example, the model would have reconstructed faces with African features. This perspective, which is common in technical AI research, treats bias as a statistical artifact that can be mitigated through better data curation—ensuring more diverse and representative datasets.

Gebru, however, strongly challenged this framing, arguing that AI bias is not merely a technical issue but a reflection of deeper societal and structural inequalities. She contended that simply diversifying datasets does not address the broader power dynamics, historical injustices, and systemic biases that shape how AI technologies are designed and deployed. From her perspective, treating bias as a technical flaw overlooks the ways in which AI systems reinforce existing social hierarchies, particularly along lines of race, gender, and economic status. This debate encapsulates a broader division within the AI community: One approach sees bias as a problem of statistical representation, while the other frames it as an issue of social justice that requires systemic change.

This tension between technical and sociopolitical perspectives on AI bias reveals the limits of purely engineering-driven solutions. While improving dataset diversity may reduce certain kinds of bias, it does not fundamentally challenge the assumptions, incentives, and power structures embedded in AI development. A more comprehensive approach to fairness in AI must therefore go beyond technical fixes to consider the broader ethical, political, and social dimensions of these technologies.

Critical thinking offers an important and complementary perspective on AI by enabling a decentering of dominant narratives about machine learning. Rather than accepting AI as a neutral or purely technical field, critical approaches interrogate the social, historical, and ideological forces that shape its development. They encourage us to ask not only how AI works but also whose interests it serves, what values it encodes, and what forms of knowledge it privileges or marginalizes. By integrating critical inquiry with technical expertise, we can move beyond narrow problem-solving frameworks and toward a deeper understanding of AI's role in shaping contemporary society. This chapter explores these themes, examining how critical perspectives enrich discussions on AI bias and open new possibilities for more just and equitable technologies.

1 Rethinking AI neutrality: power, politics, and the social shaping of technology

The public debate between Yann LeCun and Timnit Gebru over bias in machine learning brings to light a fundamental philosophical question: Is technology a neutral tool, or is it inevitably shaped by the social, political, and historical contexts in which it is created and deployed? This tension lies at the heart of debates around AI fairness and accountability, revealing a deep divide between technical explanations and broader critiques of power and social structure.

In the dominant view within computer science and engineering, AI systems are seen as neutral instruments. Problems of bias are framed as technical flaws—primarily the result of unrepresentative or skewed datasets. From this perspective, fairness can be achieved through better data and improved optimization techniques. Yet this instrumentalist conception of technology has long been challenged by philosophers and theorists who argue that such an understanding conceals the deeper ways in which technologies shape, and are shaped by, society.

Langdon Winner (2004 [1968]) warns against what he calls 'technological somnambulism'—the unreflective assumption that technologies are external to politics and values. In his account, technologies are not just passive tools but active participants in the organization of social life. Far from neutral, they embed particular arrangements of power and can reproduce or even entrench existing hierarchies (Winner, 1980). Facial recognition systems that misidentify people of color are not merely failing to perform accurately; they reflect and perpetuate long-standing racial disparities built into the institutional and social frameworks in which these tools are deployed.

This insight is central to the work of Timnit Gebru, who argues that AI systems inherit the structural inequalities of the environments in which they are developed—corporate, academic, and governmental institutions that have historically privileged certain perspectives and populations. Along with Joy Buolamwini, Gebru (2018) has shown how the very definitions of fairness and accuracy are themselves shaped by those in power. From this vantage point, technical solutions alone are insufficient because the underlying values and assumptions driving AI development remain unexamined.

The idea that technology reflects broader political and economic priorities is further developed by Andrew Feenberg (1999), who rejects technological determinism in favor of a constructivist account. For Feenberg, technological systems do not emerge in isolation from social forces; they are shaped by the intentions, interests, and constraints of those who design and control them. This means that AI development—decisions about data, model architectures, optimization goals—is inseparable from institutional agendas, particularly those of dominant corporate actors. Like Gebru, Feenberg calls for democratizing technology by opening up decision-making processes to more diverse publics, ensuring that design choices reflect not only commercial imperatives but also social and ethical concerns.

This position also aligns with broader trends in science and technology studies (STS), particularly the social construction of technology (SCOT) framework developed by Trevor Pinch and Wiebe Bijker (1984). SCOT emphasizes how different social groups influence the development and interpretation of technological artifacts. Applied to AI, this means that the priorities of large technology firms—monetization, data extraction, platform optimization—heavily shape research directions and practical deployments. Technical parameters are not neutral: They encode commercial logic and cultural assumptions that often sideline public interest or equity.

What distinguishes these perspectives is less a matter of disagreement than one of emphasis. Gebru highlights the institutional and historical reproduction of inequality; Feenberg stresses the political nature of design choices and advocates participatory governance; SCOT foregrounds the multiplicity of social actors and their negotiations in shaping technological outcomes. Together, they provide a convergent critique: AI systems are not built in a vacuum, and their so-called neutrality masks a complex interplay of power, interest, and ideology.

The illusion of technological neutrality is also challenged at a more conceptual level. Heidegger, in *The Question Concerning Technology* (1954), describes technology not merely as a means to an end but as a way of revealing the world—a mode of perception he calls *Gestell*, or enframing. From this perspective, technologies do not simply process information; they shape what is visible, sayable, and actionable. AI systems exemplify this logic: Their classifications and predictions structure how social reality is perceived and navigated. When an algorithm reconstructs a Black face as white, it is not just making a representational error—it is enacting a normative framework that excludes or distorts certain identities.

Similarly, Jacques Ellul (1964) warns that modern technological systems operate according to an autonomous logic of efficiency, subordinating ethical judgment to the pursuit of control and optimization. In AI applications, this dynamic is evident in the push for scalable, standardized solutions across domains like health care, education, and criminal justice. The result is often a disregard for complexity, difference, or moral nuance—a process that reinforces what Ellul sees as the dehumanizing trajectory of technical rationality.

What emerges from these overlapping critiques is a coherent rejection of the myth that technology—and AI in particular—is apolitical or objective. The debate between LeCun and Gebru thus represents more than a technical disagreement: It marks a deeper philosophical divide over whether technologies can be disentangled from the power structures that shape them. Understanding AI through this lens requires more than refining datasets or improving metrics. It calls for a sustained interrogation of the institutions, values, and interests that determine how technologies are built, evaluated, and used.

Addressing bias in AI is not just a matter of technical correction but one of political imagination. It demands new forms of governance, broader participation in technological design, and a willingness to confront the histories of exclusion and domination that continue to inform the digital infrastructures of the present.

2 Decentering the AI perspective through critical thinking

Critical thinking, broadly defined, refers to a reflective and questioning attitude that challenges assumptions, examines power relations, and evaluates the ethical consequences of decisions and practices. In the context of AI, critical thinking moves beyond narrow technical fixes to address broader issues of power, agency, and epistemology. It asks not only how AI works but whose interests it serves, what values it encodes, and who might be marginalized or disadvantaged by its deployment.

2.1 Critical thinking and the situated ethics of fairness

This approach is essential for confronting the structural biases and inequities that AI systems can reproduce. Universal values such as fairness and equality often motivate efforts to improve AI, but if these values are applied in a supposedly neutral, context-free manner they risk reproducing the very power structures they aim to overcome (Noble, 2018; Benjamin, 2019). AI-systems trained on data reflecting systemic inequalities tends to amplify those same patterns, even under the banner of objectivity. This problem has been extensively documented in AI ethics literature, especially for image datasets (Birhane & Prabhu, 2021).

Fairness itself is not a singular, universal principle but a contested and situated concept (Binns, 2018). Different cultural and philosophical traditions define fairness differently, whether in terms of equal treatment or equity that accounts for structural disadvantage. Critical thinking helps expose these differences and challenges the tendency of AI to adopt dominant cultural standards as if they were universal norms.

These competing definitions of fairness do not exist in a vacuum. They are mediated by institutions and corporations that hold the power to standardize values in technical systems. As Kate Crawford (2021) has argued, the global influence of major technology companies often imposes particular epistemologies and values on a worldwide scale, reinforcing existing hierarchies rather than dismantling them. Critical thinking is what allows us to interrogate how such power dynamics shape the very categories of fairness that are operationalized in AI.

Although sometimes dismissed as impractical, critical thinking is indispensable. It complements policy reform, interdisciplinary collaboration, and participation by affected communities. By continually questioning assumptions, it supports the development of AI systems that are not only technically robust but also ethically and socially accountable. In this sense, critical thinking is not opposed to universal values but helps to ensure they remain inclusive, dynamic, and open to revision, strengthening their legitimacy and relevance in a diverse world.

2.2 Lessons from the stochastic parrot debate

A second important critical perspective emerges from the 'stochastic parrot' debate, as named by Bender et al. (2021), which has drawn significant attention to the limits and risks of LLMs. As discussed in Chapter 1, Bender and colleagues argue that LLMs generate language based on probabilistic patterns without semantic or conceptual understanding, making them prone to several well-documented harms.

First, these models can reproduce and amplify social biases present in their training data, entrenching harmful stereotypes (Gururangan et al., 2022). Second, their enormous computational demands have severe environmental impacts, raising questions about responsible resource use and sustainability (Strubell et al., 2019)—we will come back to this in Chapter 9. Third, their capacity to generate vast amounts of synthetic, plausible-sounding text without verification mechanisms creates a fertile ground for misinformation, deepfakes, and the erosion of public trust in communication (Floridi and Chiriatti, 2020). These issues align with broader philosophical critiques about how AI systems risk reinforcing existing structures of domination rather than challenging them.

Beyond these specific risks, there is a deeper epistemic concern: By outsourcing language production to statistical models, society risks diminishing the role of human agency in meaning-making, potentially standardizing and narrowing discourse (Foucault, 1971). This threatens to undermine creativity and critical engagement, especially in fields like journalism, education, or the arts.

These challenges are not merely theoretical. Real-world incidents—biased hiring algorithms, AI-generated disinformation, or discriminatory surveillance—demonstrate the tangible consequences of deploying LLMs without adequate oversight. Addressing these concerns requires a multidisciplinary effort that bridges technical, philosophical, and sociopolitical analysis. The central question remains how to harness the potential of LLMs while aligning their development with social justice, democratic accountability, and ecological responsibility.

3 AI bias, universalism, and the epistemology of representation

While often approached as a purely technical challenge, bias in AI runs much deeper: It is intrinsic to any system that attempts to represent the world. Any act of representation—whether in language, images, or statistical models—inevitably involves a perspective, a framing of reality according to certain choices, norms, and assumptions. This fundamental fact creates a philosophical tension: If universalism of values demands neutrality, objectivity, and fairness, how can it be reconciled with the inevitability of bias? This tension is particularly pressing in AI ethics, where the goal of creating universally fair, unbiased systems conflicts with the reality that any AI system necessarily reflects a particular point of view.

To represent the world—even in statistical or linguistic form— is to make selections, frame priorities, and encode normative assumptions. In this sense, LLMs do not merely mirror the world; they inherit, reproduce, and often amplify the historical, cultural, and political structures embedded in their training data. When they generate language, they enact a kind of epistemic performance—one that reflects not only patterns in data but the social imaginaries and power relations from which those patterns emerged.

The dominant epistemology of AI—especially in mainstream computer science—tends to assume that meaning is reducible to patterns of use and that knowledge can be generalized across contexts through statistical regularities. This model leaves little room for the kinds of normative contestation, positionality, and epistemic pluralism that characterize human life. It also risks universalizing a particular dominant standpoint, under the guise of objectivity. As scholars in STS have argued, what is presented as neutral computation is often 'a view from nowhere' that masks its own partiality (Haraway, 1988; Nagel, 1986).

The task, then, is not simply to 'debias' AI but to critically interrogate the epistemological foundations of AI systems: how they encode knowledge, whose knowledge they prioritize, and what forms of knowing are marginalized or excluded. Rather than viewing bias as a removable error, this chapter argues that bias is a structural condition of representation and an unavoidable feature of systems that are situated within specific social and historical contexts. This view calls for a reframing of fairness not as an ideal state of neutrality but as an ongoing negotiation among competing normative commitments. The sections that follow will develop this argument through a typology of AI biases and a series of critical perspectives—feminist, decolonial, and antiracist—that illuminate the political and ethical stakes of how LLMs know and speak.

3.1 A typology of biases in AI systems

A more nuanced understanding of bias requires moving beyond the singular term 'bias' to a pluralistic and structured typology. Not all biases in AI systems are of the same kind, and not all arise from the same causes or lead to the same consequences. As Suresh and Guttag (2021) argue, different biases operate at different stages of the machine learning pipeline, from data collection to model deployment. A typology of bias thus serves both a diagnostic and philosophical function: It clarifies the different ways in which AI systems can become entangled with social inequalities and invites deeper reflection on the epistemic assumptions underlying these systems. The typology below follows closely the one proposed by Suresh and Guttag (2021):

Historical bias. Historical bias occurs when the data used to train AI systems reflects preexisting structural inequalities, exclusions, or injustices. This form of bias is not only a product of poor data sampling or technical error but a consequence of faithfully mirroring a world that is already unjust. For example, predictive policing algorithms that rely on past arrest data risk reinforcing patterns of racial profiling, simply because historical data reflects discriminatory policing practices (Richardson et al., 2019).

Philosophically, historical bias raises the question of whether knowledge that accurately reflects the past can still be unjust. As Hacking (1990) has argued, statistical representation is always entangled with the categories through which institutions intervene in the world; the feedback loop between representation and reality means that AI systems do not merely describe the world but actively participate in its reproduction.

Representation bias. Representation bias arises when certain groups or perspectives are underrepresented or misrepresented in training datasets. In the case of LLMs, this can occur when the corpus consists predominantly of text from demographic majorities—often male, white, and from the Global North—leading to models that systematically neglect or stereotype other voices. This issue is particularly salient for systems that claim general-purpose applicability, as their universality is undermined by their parochial training base.

Representation bias also intersects with epistemic justice. Fricker's (2007) concept of testimonial injustice—the unfair devaluation of a speaker's credibility based on identity—helps explain why some perspectives are systematically excluded from training corpora or treated as less authoritative. LLMs trained on web data inherit these exclusions and may unwittingly reproduce them in their outputs.

Measurement bias. Measurement bias occurs when the features or proxies used in modeling do not capture the underlying phenomena in an equitable or meaningful way. For instance, using zip code as a proxy for creditworthiness can encode socioeconomic and racial segregation into ostensibly neutral models. In LLMs, sentiment analysis tools trained on specific dialects may fail to detect sarcasm or cultural nuance, leading to misclassifications that reflect normative assumptions about language use.

This form of bias highlights the epistemological problem of proxy variables—that is, the reduction of complex social constructs to quantifiable metrics. It echoes Hacking's (1986) critique of 'making up people': the ways in which categories and classifications shape the realities they purport to measure.

Aggregation bias. Aggregation bias results from the assumption that a single model can be equally valid for all users or subpopulations. This is particularly problematic in settings where group differences are significant—such as health, education, or language use. Models that generalize across diverse populations often perform best for majority groups and worst for marginalized ones, effectively encoding inequity through statistical averaging.

The ethical issue here is not only distributive (who gets a worse prediction) but epistemic: Whose variation is considered noise and whose is modeled as signal? Aggregation bias illustrates the philosophical limits of universalist assumptions in machine learning—particularly the presumption that fairness can be defined independently of social context.

Deployment bias. Even well-designed models can generate harm if deployed in contexts for which they were not intended. Deployment bias refers to the gap between a model's training environment and its real-world application. For instance, a language model fine-tuned for summarizing news articles might be used to assess legal documents or mental health records, with unpredictable consequences. This form of bias underscores the importance of situated ethics and contextual evaluation.

It also reflects broader concerns in philosophy of technology about technological determinism and the social construction of use (Oudshoorn and Pinch, 2003; Winner, 1980). Tools are not neutral; their effects depend on how they are integrated into human practices and institutional logics.

Epistemic and normative bias. Finally, epistemic bias refers to the normative assumptions about knowledge, truth, and value that are baked into the architecture of AI systems. LLMs, for instance, often rely on statistical frequency as a proxy for salience or credibility, effectively equating commonness with correctness. This creates what we might call a majoritarian epistemology, where dominant perspectives are amplified and dissenting ones are marginalized.

This issue resonates with critiques of algorithmic normativity—the ways in which statistical regularity comes to stand in for moral or epistemic authority. As Longino (1990) has emphasized in feminist epistemology, objectivity requires the inclusion of diverse perspectives, not their erasure under the guise of universalism.

These forms of bias are not mutually exclusive, and in practice they often reinforce one another. What binds them together is their epistemological significance: They reveal that AI systems are not merely tools for representing the world but participants in the ongoing construction of social reality. While Suresh and Guttag (2021) offer a comprehensive typology rooted in the technical pipeline of machine learning systems, Hovy and Prabhumoye (2021) propose a slightly different classification, one grounded more explicitly in the linguistic and social dimensions of bias in natural language processing (NLP) (biases linked to data, annotation, input representations, algorithms/models, or research design/evaluation). The next sections will explore how different critical traditions—feminist, antiracist, decolonial—respond to this condition, offering alternative ways of understanding fairness, objectivity, and knowledge in the age of LLMs.

3.2 Feminist approaches: gender bias and situated knowledges

Feminist theory offers one of the most sustained critiques of the ideal of objectivity that underlies much mainstream AI discourse. Instead of viewing bias as a deviation from a neutral norm, feminist epistemologies understand all knowledge as situated, partial, and shaped by power. LLMs, from this perspective, are not neutral mirrors but discursive agents trained on culturally specific, often patriarchal corpora. Their outputs reflect not only statistical patterns but also the social imaginaries and institutional power structures that shape those patterns.

Donna Haraway's concept of situated knowledge (1988) is foundational in this context. Rather than aspiring to a 'God's eye view,' Haraway argues that objectivity requires acknowledging and accounting for one's position. Applied to AI, this view suggests that LLMs should not be judged by illusory standards of epistemic neutrality but by the standpoints they reproduce and the exclusions they enact. When LLMs generate language, they do so within a matrix of learned associations—many of which naturalize dominant, often exclusionary, social norms.

One well-documented consequence is the reproduction of gender stereotypes in NLP systems. Bolukbasi et al. (2016) show that word embeddings associate 'man' with 'programmer' and 'woman' with 'homemaker,' revealing how models trained on large corpora can amplify and legitimize systemic inequality. These outputs contribute to what Fricker (2007) calls 'hermeneutical injustice': the structural harm that arises when marginalized groups lack the conceptual resources to articulate or validate their experiences. In this sense, LLMs not only reflect social bias but participate in its epistemic reproduction.

Feminist critiques thereby expand the notion of bias from technical error to epistemic violence—the silencing, distortion, or devaluation of certain ways of knowing. This perspective aligns with critiques of AI systems as 'technologies of classification' (Bowker and Star, 1999), where algorithmic outputs structure the categories through which people are recognized and governed.

Intersectionality, a concept introduced by Kimberlé Crenshaw, further deepens this analysis by examining how systems of oppression—such as racism, sexism, and classism—interact in compounding ways. An intersectional approach to AI highlights how models trained on unmarked or majority data often misrepresent or exclude those at the margins, including women of color, trans individuals, and disabled users. The consequences are not only representational. Studies such as that by Buolamwini and Gebru (2018) on racial and gender disparities in facial recognition show that technical failures can translate into social harm. In the case of LLMs, underrepresentation of dialects like African American Vernacular English can lead to misclassification in sentiment analysis systems (Sap et al., 2019), reinforcing harmful stereotypes and producing material consequences in domains like hiring, policing, and moderation.

In response, feminist theorists have advanced design methodologies grounded in reflexivity, care, and participatory justice. As Costanza-Chock (2020) argues in *Design Justice*, AI systems that are more just must center the lived experiences of marginalized communities from the outset, rather than retrofitting fairness after the fact. These approaches echo value-sensitive design (Friedman and Hendry, 2019) and support participatory machine learning models in which affected communities are not merely data sources but epistemic agents in shaping technological futures.

Such methodologies rest on relational theories of knowledge, emphasizing mutual accountability and recognition over abstraction and formal rigor. They challenge dominant models of AI development that often prioritize efficiency and generalizability at the expense of context and specificity.

Finally, feminist theorists have drawn attention to the politics of voice in AI systems. While LLMs are often evaluated in terms of fluency and coherence, the deeper question is one of audibility: Whose voices are recognized as legitimate? When models privilege dominant linguistic norms and suppress or exoticize others, they risk contributing to forms of epistemic silencing, in the sense that Spivak (1988) critiques in 'Can the Subaltern Speak?' The question she poses—'Can the subaltern speak?'—might be reframed today, in the context of generative AI, as 'Can the subaltern be generated?'

This is not merely a rhetorical question but a demand for epistemic responsibility. It challenges designers, researchers, and institutions to consider how language models might be made responsive to the multiplicity of voices and lifeworlds they purport to represent. Feminist approaches thus reframe AI bias not simply as a technical flaw but as a symptom of structural inequality and epistemic exclusion. They call for new ways of knowing—grounded in positionality, accountability, and justice.

3.3 Race, power, and critical algorithm studies

Previous discussions have shown that racial bias in AI cannot be reduced to technical failures alone but must be understood within broader patterns of power and exclusion. Benjamin (2019) offers a particularly incisive account of this dynamic through the concept of the 'New Jim Code'—a regime in which ostensibly neutral algorithms sustain the logics of racial segregation and inequality. Just as segregationist laws in the United States (the Jim Crow laws) once enforced racial subordination, algorithmic systems today influence who is surveilled, who is deemed risky or credible, and whose speech is amplified or ignored. LLMs trained on vast internet corpora frequently inherit racially encoded language, including stereotypes, slurs, and skewed associations. Studies show that such models disproportionately associate Black-sounding names with negative sentiment or lower-status occupations (Wilson and Caliskan, 2024), thereby reinforcing a symbolic economy in which certain identities are rendered suspect or disposable. These are not mere linguistic missteps—they contribute to what Mills (1997) describes as an 'epistemology of ignorance,' whereby dominant groups sustain their authority through the erasure or distortion of marginalized experiences.

This erasure extends to questions of epistemic authority. Decisions about what counts as fair, accurate, or neutral knowledge are often made by institutions—tech companies, elite research labs, regulatory agencies—that are demographically and ideologically narrow. As Collins (1990) has shown, dominant knowledge practices reinforce a 'matrix of domination' in which race and gender hierarchies persist through control over naming, classification, and legitimacy. In the case of LLMs, this is evident in how alternative linguistic traditions (see previous section) are frequently misinterpreted or pathologized by models calibrated to white-coded norms. Such outcomes are not only unjust; they constitute forms of epistemic violence, denying legitimacy to culturally specific modes of expression and understanding.

Confronting these issues requires more than technical audits or fairness metrics. The movement for algorithmic justice, led by organizations such as Data for Black Lives and the Algorithmic Justice League, calls for structural transformation rooted in abolitionist and decolonial frameworks. Abolition, in this context, does not entail the rejection of technology per se, but of systems that reproduce carceral logics, epistemic domination, and exploitative labor. It entails redistributing epistemic power: shifting the terms of design, evaluation, and governance to include those most affected by AI systems. This includes practices such as community-led audits, data sovereignty initiatives, and epistemic reparations—efforts to recognize and restore marginalized knowledge traditions.

Philosophically, these interventions resonate with decolonial thought, which challenges the ideal of universal, disembodied knowledge in favor of pluralism, relationality, and context. Scholars such as Mignolo (2011) and Wynter (2003) argue that decoloniality requires a fundamental rethinking of who gets to know and what is considered knowledge. In the domain of LLMs, this means developing models that do not simply perform well by dominant standards but that are accountable to diverse epistemic communities and histories.

In sum, racial bias in language models is not an isolated defect but a symptom of entrenched systems of exclusion, exploitation, and epistemic violence. Addressing it demands not only better data or algorithms but a reconfiguration of the political, economic, and epistemological assumptions that structure AI development. The following section will explore how these dynamics unfold in global and multilingual contexts, where linguistic hierarchies and geopolitical asymmetries further shape the reach and relevance of LLMs.

3.4 Cultural and linguistic biases: epistemic coloniality

Cultural and linguistic biases in LLMs are often discussed in terms of fairness or inclusivity, but such frameworks risk obscuring the deeper global asymmetries that shape contemporary knowledge production. Most LLMs are trained predominantly on English-language data, sourced disproportionately from the Global North and filtered through infrastructures designed within Euro-American epistemic frameworks. This can be seen not merely as a reflection of a lack of diversity but as a reproduction of epistemic coloniality—the systemic privileging of certain languages, cultures, and worldviews over others.

This dynamic is part of a longer history in which colonialism functioned not only as a political and economic project but also as an epistemic one. It imposed a hierarchical ontology that framed Western rationality as universal and relegated other knowledge systems to the margins. Today, LLMs risk perpetuating this legacy under the guise of computational objectivity and global scalability. Their architectures, data pipelines, and optimization goals often reflect dominant linguistic and cultural norms, flattening the pluralism of human expression into statistically average outputs aligned with hegemonic standards.

Linguistic hegemony remains one of the most persistent forms of bias, with English and other high-resource languages vastly overrepresented in training and evaluation (for example, in the case of Llama 3, approximately 95% of the training data used by Meta was in English; the notion of 'other high-resource languages' should therefore be put into perspective, as they collectively account for less than 5% of the data).[1] While recent advances in multilingual modeling have improved performance in some additional languages, low-resource languages—often spoken by marginalized communities—remain underrepresented in data, research investment, and infrastructure. As Spivak (1988) argues in her formulation of the subaltern condition, when certain voices cannot be represented or understood within dominant discursive frames, their silencing is not incidental but structural.

The aspiration to build 'universal' language models often masks this asymmetry. Universality, in practice, tends to mean standardization according to dominant norms rather than a genuinely pluralistic architecture capable of engaging epistemic difference. As Appiah (2006) suggests, cosmopolitanism must be rooted—valuing difference within connection—rather than serving as a universalizing project that, as some critics argue, risks becoming a form of 'imperial cosmopolitanism' that conflates inclusion with control.

---

[1] Source: https://ai.meta.com/blog/meta-llama-3.

Cultural bias in LLMs extends beyond language to encompass ontological assumptions embedded in knowledge representation systems—taxonomies, classification schemas, and moderation filters—that often reproduce Anglo-European categories. Entire knowledge domains, such as Indigenous kinship systems or local ecological practices, may be misrepresented or erased, not because of technical failure but because they fall outside the epistemic boundaries recognized by dominant datasets. Boaventura de Sousa Santos (2013) calls this epistemicide: the systemic devaluation or destruction of nondominant knowledge systems.

These issues also surface in practices of translation, where LLMs often reduce culturally rich expressions to functional equivalents intelligible within dominant languages. From a philosophical and cultural standpoint, translation is not merely a technical act but an ethical and political negotiation between worldviews. As Venuti (2008) argues, dominant translation practices frequently domesticate the foreign—rendering culturally specific content in terms that conform to the expectations of the target culture. When sacred or culturally embedded concepts are mapped onto generic or Western-coded terms, the result can be a form of erasure or distortion, with real social and epistemic consequences. In contexts such as global education, digital assistance, or intercultural communication tools, such translations risk reinforcing normative hierarchies while presenting themselves as neutral solutions to cultural difference.

Efforts to mitigate these harms require more than technical refinements; they call for epistemic humility—an awareness of what models cannot know or adequately represent. This entails designing systems that foreground contestability and allow users to contextualize or challenge outputs. Such approaches reimagine LLMs not as authoritative speakers but as dialogical partners embedded in relations of learning, care, and negotiation.

These concerns are inseparable from the geopolitics of AI development. The material and institutional infrastructures behind LLMs—computing resources, data pipelines, research

funding—are concentrated in a few powerful countries and corporations (we will come back to this in Chapter 9). As a result,
decisions about representation, safety, and appropriateness are
often made without meaningful participation from those most
affected. While international efforts such as UNESCO's Recommendation on the Ethics of AI (2021) have acknowledged the
need for cultural adaptability, implementation remains uneven,
and the risk of normative imperialism persists.

In response, movements advocating for AI and data sovereignty have emerged across the Global South and among Indigenous communities. These efforts seek to reclaim control over
how cultural knowledge is represented and governed, emphasizing relational understandings of data and accountability to specific moral and ecological contexts (Carroll et al., 2019). From
this perspective, data is not a universal resource but a situated
good, entangled with questions of consent, responsibility, and
cultural survival.

In sum, the cultural and linguistic biases embedded in LLMs are
not technical oversights but symptoms of deeper epistemic hierarchies and colonial legacies. Addressing them requires a rethinking of foundational assumptions in language modeling, including
how we define knowledge, whose voices are centered, and what
forms of difference our models are capable of respecting. The next
section turns to possible futures for LLMs grounded in pluralism,
reflexivity, and epistemic justice.

3.5 Pluralism, reflexivity, and epistemic justice

If bias in LLMs is not a peripheral malfunction but a structural
feature of how knowledge is encoded, then addressing it demands
more than improved data or refined metrics. It calls for a reorientation in the epistemological foundations of AI—toward systems
that do not claim neutrality but actively engage with epistemic
pluralism, reflexivity, and justice. As previous sections have
shown, gender, race, and culture are not merely dimensions of

representational diversity but deep entanglements that shape what counts as knowledge, whose language is intelligible, and whose values are built into algorithmic infrastructures.

Mainstream approaches to fairness in AI tend to rely on formal definitions—statistical parity, equalized odds, or counterfactual reasoning—that often abstract away from the historical and social conditions in which these models operate. While such frameworks help detect quantifiable harms, they frequently mask the deeper question: fairness according to whom, and in service of which vision of the world? Procedural metrics, while necessary, are insufficient; they risk reducing justice to distribution, rather than addressing questions of recognition, legitimacy, and epistemic authority. This challenges liberal universalist models of fairness and invites a pluralist ethics capable of engaging conflicting yet equally valid modes of knowing.

The concept of epistemic justice, particularly as theorized by Fricker (2007), provides a compelling lens here. Her distinction between testimonial injustice—when credibility is unfairly denied to a speaker due to prejudice—and hermeneutical injustice (see Section 3.2)—maps closely onto the risks posed by LLMs. These models do not merely reproduce dominant discourses; they also help shape the conceptual boundaries of intelligibility, structuring whose voices are heard and whose remain inaudible. Achieving epistemic justice requires more than representational inclusion; it entails a transformation in how systems are designed and governed, beginning with a commitment to participatory epistemology—where affected communities are not only sources of data but cocreators of meaning.

Central to this reimagining is the principle of reflexivity. Informed by feminist science studies and critical design theory, reflexivity entails not just transparency but a system's (and designer's) capacity for self-awareness and critical responsiveness to its own positionality. This could mean designing models that acknowledge uncertainty, foreground gaps in training data, or reveal conflicting outputs. It also entails moving from product-driven notions of AI to processual ones, where models evolve through iterative engagement, ethical dialogue, and ongoing negotiation with diverse publics. In this sense, LLMs should function less as authoritative answer-machines and more as tools for collaborative interpretation—supporting the dialogical process of making meaning rather than foreclosing it.

Such an orientation also speaks to a broader philosophical tension between universalism and relativism in AI ethics. Attempts to impose global standards risk reproducing the very forms of epistemic dominance they seek to address, while radical contextualism can undermine shared norms and accountability. A more promising alternative is relational universalism: a mode of ethical reasoning that affirms the value of universal principles but insists they must be interpreted through, and accountable to, the situated experiences of diverse communities. Drawing on hermeneutic ethics (Ricoeur, 1992), feminist standpoint theory (Harding, 1991), and decolonial humanism (Wynter, 2003), this view reconfigures universality as an ongoing process of normative negotiation, rather than a fixed set of rules.

Designing for such pluralism requires institutional support. Epistemic justice cannot be achieved through model architecture alone; it depends on governance structures that redistribute epistemic authority and embed mechanisms for accountability. These might include community oversight boards to audit training practices, data cooperatives that allow collective control over linguistic and cultural representation, and systems of contestability that empower users to challenge or annotate outputs. Support for nondominant languages and grassroots AI initiatives is likewise essential if we are to counter the epistemic concentration of power in a handful of institutions and languages.

Confronting the biases of LLMs ultimately requires a shift in how we conceive of language technologies: not as neutral instruments of communication but as political and epistemic agents that shape how the world is understood and acted upon. What is at stake is not only the performance of models but the values they instantiate, the worlds they render intelligible, and the futures they help imagine. Epistemic justice is thus not a technical destination but a continuous practice of inquiry, recognition, and repair—an open-ended task of designing AI systems that can live with, and learn from, the plurality of human thought.

4 Conclusion: reimagining fairness through situated universals

If bias in AI cannot be fully eliminated, then the task before us is not to perfect neutrality but to reimagine what fairness might mean in a world of irreducible difference. Throughout this chapter, we have seen that bias in LLMs is not merely a statistical distortion or technical flaw. It is a mirror—sometimes distorting, sometimes clarifying—of deeper epistemic structures: who has the power to define knowledge, whose language counts as normative, and which values guide systems that increasingly mediate human understanding.

Against the ideal of a universal fairness standard, we have proposed a model of relational universalism: an ethical stance that affirms shared principles—such as dignity, accountability, and justice—while recognizing that these must be interpreted through the lens of context, history, and situated experience. Fairness, in this view, is not uniformity but just difference: an openness to plural ways of knowing, speaking, and being, without abandoning the need for mutual recognition and collective responsibility.

This requires embracing the paradox at the heart of pluralist ethics: that universality is most meaningful not when it erases difference but when it is continually negotiated across it. It is not a case of unbounded relativism but rather one of ethical reflection informed by principles such as care, humility, and accountability. In practice, this means designing AI systems that are reflexive, contestable, and shaped through inclusive governance. It means moving beyond metrics to ask more fundamental questions: Whose voices are heard? Whose perspectives are encoded? What futures are being made possible—or foreclosed—by the models we build?

Reimagining fairness in this way does not simplify the problem of bias; it deepens it. But, in doing so, it also expands the horizon of what ethical AI could become: not a neutral arbiter of information but a participant in the ongoing, collective work of justice.

References

Appiah, K. A. (2006). *Cosmopolitanism: Ethics in a world of strangers.* W. W. Norton & Company.

Bender, E. M., Gebru, T., McMillan-Major, A., and Shmitchell, S. (2021). On the dangers of stochastic parrots: Can language models be too big? 🦜. In *Proceedings of the 2021 ACM Conference on Fairness, Accountability, and Transparency (FAccT '21), New York, NY, USA* (pp. 610–623). Association for Computing Machinery. https://doi .org/10.1145/3442188.3445922

Benjamin, R. (2019). *Race after technology: Abolitionist tools for the New Jim Code.* Polity Press.

Binns, R. (2018). Fairness in machine learning: Lessons from political philosophy. In *Proceedings of the 1st Conference on Fairness, Accountability and Transparency (FAC), New York, USA, PMLR, 81,* 149–159. https://proceedings.mlr.press/v81/binns18a.html

Birhane, A., and Prabhu, V. U. (2021). Large image datasets: A pyrrhic win for computer vision? In *2021 IEEE Winter Conference on Applications of Computer Vision (WACV), Waikoloa, HI, USA, 2021* (pp. 1536–1546). https://doi.org/10.1109/WACV48630.2021.00158

Bolukbasi, T., Chang, K., Zou, J., Saligrama, V., and Kalai, A. (2016). Man is to computer programmer as woman is to homemaker? Debiasing word embeddings. In *Proceedings of the 30th International Conference on Neural Information Processing Systems (NIPS'16), Red Hook, NY, USA* (pp. 4356–4364). Curran. https://dl.acm.org /doi/10.5555/3157382.3157584

Bowker, G. C., and Star, S. L. (1999). *Sorting things out: Classification and its consequences.* The MIT Press.

Buolamwini, J., and Gebru, T. (2018). Gender shades: Intersectional accuracy disparities in commercial gender classification. In S. A. Friedler and C. Wilson (Eds.), *Proceedings of the 1st Conference on Fairness, Accountability and Transparency (FAC), New York, USA* (pp. 77–91). PMLR. https://proceedings.mlr.press/v81/buolamwini18a /buolamwini18a.pdf

Carroll, S. R., Rodriguez-Lonebear, D., and Martinez, A. (2019). Indigenous data governance: Strategies from United States native nations. *Data Science Journal*, 18, 31. https://doi.org/10.5334/dsj-2019-031

Collins, P. H. (1990). *Black feminist thought: Knowledge, consciousness, and the politics of empowerment*. Routledge.

Costanza-Chock, S. (2020). *Design justice: Community-led practices to build the worlds we need*. The MIT Press.

Crawford, K. (2021). *Atlas of AI: Power, politics, and the planetary costs of artificial intelligence*. Yale University Press.

Ellul, J. (1964). *The technological society* (J. Wilkinson, Trans.). Vintage Books.

Feenberg, A. (1999). *Questioning technology*. Routledge.

Floridi, L., and Chiriatti, M. (2020). GPT-3: Its nature, scope, limits, and consequences. *Minds and Machines*, 30(4), 681–694. https://doi.org/10.1007/s11023-020-09548-1

Foucault, M. (1971). Orders of discourse. *Social Science Information*, 10(2), 7–30. https://doi.org/10.1177/053901847101000201

Fricker, M. (2007). *Epistemic injustice: Power and the ethics of knowing*. Oxford University Press.

Friedman, B., and Hendry, D. G. (2019). *Value sensitive design: Shaping technology with moral imagination*. The MIT Press.

Gururangan, S., Card, D., Dreier, S., Gade, E., Wang, L., Wang, Z., Zettlemoyer, L., and Smith, N. A. (2022). Whose language counts as high quality? Measuring language ideologies in text data selection. In *Proceedings of the 2022 Conference on Empirical Methods in Natural Language Processing (EMNLP), Abu Dhabi, United Arab Emirates* (pp. 2562–2580). Association for Computational Linguistics. https://aclanthology.org/2022.emnlp-main.165/

Hacking, I. (1986). Making up people. In T. C. Heller, M. Sosna, and D. E. Wellbery (Eds.), *Reconstructing individualism: Autonomy, individuality, and the self in Western thought* (pp. 222–236). Stanford University Press.

Hacking, I. (1990). *The taming of chance*. Cambridge University Press.

Haraway, D. J. (1988). Situated knowledges: The science question in feminism and the privilege of partial perspective. *Feminist Studies*, 14(3), 575–599. https://doi.org/10.2307/3178066

Harding, S. (1991). *Whose science? Whose knowledge? Thinking from women's lives*. Cornell University Press.

Heidegger, M. (1977 [1954]). *The question concerning technology and other essays* (W. Lovitt, Trans.). Harper & Row.

Hovy, D., and Prabhumoye, S. (2021). Five sources of bias in natural language processing. *Language and Linguistic Compass, 15*(8), e12432. https://doi.org/10.1111/lnc3.12432

Longino, H. E. (1990). *Science as social knowledge: Values and objectivity in scientific inquiry*. Princeton University Press.

Mignolo, W. D. (2011). *The darker side of Western modernity: Global futures, decolonial options*. Duke University Press.

Mills, C. W. (1997). *The racial contract*. Cornell University Press.

Nagel, T. (1986). *The view from nowhere*. Oxford University Press.

Noble, S. U. (2018). *Algorithms of oppression: How search engines reinforce racism*. New York University Press.

Oudshoorn, N., and Pinch, T. (Eds.). (2003). *How users matter: The co-construction of users and technology*. MIT Press.

Pinch, T. J. and Bijker, W. E. (1984). The social construction of facts and artifacts: Or how the sociology of science and the sociology of technology might benefit each other. *Social Studies of Science, 14*(3), 399–441. https://doi.org/10.1177/030631284014003004

Richardson, R., Schultz, J. M., and Crawford, K. (2019). Dirty data, bad predictions: How civil rights violations impact police data, predictive policing systems, and justice. *New York University Law Review Online, 94*, 192–233.

Ricoeur, P. (1992). *Oneself as another* (K. Blamey, Trans.). University of Chicago Press.

Sap, M., Card, D., Gabriel, S., Choi, Y., and Smith, N. A. (2019). The risk of racial bias in hate speech detection. In *Proceedings of the 57th Annual Meeting of the Association for Computational Linguistics (ACL), Florence, Italy* (pp. 1668–1678). Association for Computational Linguistics. https://doi.org/10.18653/v1/P19-1163

Santos, B. de S. (2013). *Epistemologies of the South: Justice against epistemicide*. Routledge.

Spivak, G. C. (1988). Can the subaltern speak? In: C. Nelson and L. Grossberg (Eds.), *Marxism and the interpretation of culture* (pp. 271–313). University of Illinois Press.

Strubell, E., Ganesh, A., and McCallum, A. (2019). Energy and policy considerations for deep learning in NLP. In *Proceedings of the 57th Annual Meeting of the Association for Computational Linguistics (ACL), Florence, Italy* (pp. 3645–3650). Association for Computational Linguistics. https://aclanthology.org/P19-1355/

Suresh, H., and Guttag, J. V. (2021). A framework for understanding sources of harm throughout the machine learning life cycle. *In Proceedings of the 2021 ACM Conference on Equity and Access in Algorithms, Mechanisms, and Optimization (EAAMO '21)*. Association for Computing Machinery. https://doi.org/10.1145/3465416.3483305

UNESCO. (2021). *Recommendation on the ethics of artificial intelligence*. United Nations Educational, Scientific and Cultural Organization. https://unesdoc.unesco.org/ark:/48223/pf0000381137

Venuti, L. (2008). *The translator's invisibility: A history of translation* (2nd ed.). Routledge.

Wilson, K., and Caliskan, A. (2024). Gender, race, and intersectional bias in resume screening via language model retrieval. In *Proceedings of the 2024 AAAI/ACM Conference on AI, Ethics, and Society* (pp. 1578–1590). AAAI Press. https://dl.acm.org/doi/10.5555/3716662.3716799

Winner, L. (1980). Do artifacts have politics? *Daedalus*, 109(1), 121–136.

Winner, L. (2004 [1968]). Technologies as forms of life. In David M. Kaplan (Ed.), *Readings in the philosophy of technology*. Rowman & Littlefield.

Wynter, S. (2003). Unsettling the coloniality of being/power/truth/freedom: Towards the human, after man, its overrepresentation—An argument. *CR: The New Centennial Review*, 3, 257–337.

CHAPTER 8

Disinformation, misinformation, and the crisis of trust in AI-generated content

The proliferation of large language models (LLMs) has intensified concerns about misinformation, disinformation, and the broader epistemic challenges of AI-generated content. While bias in AI primarily concerns structural inequalities and power dynamics in knowledge production, misinformation and disinformation raise additional questions about truth, manipulation, and trust in the information ecosystem (Floridi, 2014; Zannettou et al., 2019). The ability of LLMs to generate coherent, plausible, and often highly persuasive text at scale introduces a fundamental shift in the dynamics of information dissemination, blurring the boundaries between truth and fiction, authenticity and fabrication, and knowledge and propaganda.

A striking example emerged in 2022 when researchers identified networks of social media accounts disseminating articles, blog posts, and even fabricated eyewitness testimonies supporting the Russian government's version of events. These texts, often indistinguishable from legitimate news reports, portrayed Ukraine as

How to cite this book chapter:
Poibeau, T. 2025. *Understanding Conversational AI: Philosophy, Ethics and Social Impact of Large Language Models*. Pp. 177–202. London: Ubiquity Press. DOI: https://doi.org/10.5334/bde.i. License: CC BY-NC 4.0

an aggressor, fabricated evidence of Western conspiracies, and undermined reports of Russian war crimes. While such content was most likely human-authored and amplified through bots, it foreshadowed the later use of large language models to automate disinformation at scale.

A concerning feedback loop has emerged: Troll farms, saturated with fake news, are themselves captured by web crawlers that feed LLMs, subtly influencing their responses to key questions. Studies have shown that some chatbots unknowingly cite propaganda sources, reproducing false narratives as facts. This process is exacerbated by efforts to insert disinformation into publicly accessible knowledge sources, such as Wikipedia, further shaping the data LLMs rely on. While there is no definitive proof that propaganda networks deliberately optimize content for AI training, their vast and persistent output increases the likelihood of infiltration. As LLMs become more integral to information ecosystems, they risk being manipulated into unwitting amplifiers of state-sponsored disinformation.

This case illustrates how LLMs not only accelerate the spread of disinformation but also reshape its nature. The plausibility and fluency of AI-generated text blur the line between propaganda and independent journalism, creating an epistemic environment where truth becomes harder to verify and disinformation becomes more insidious. This chapter explores the role of LLMs in the production and amplification of misinformation and disinformation, examining the philosophical and political implications of automated knowledge generation. It argues that the challenges posed by AI-generated misinformation are not merely technical but deeply epistemological and political, requiring critical inquiry into the very nature of knowledge, truth, and authority in the digital age.

1 The epistemology of AI-generated content: between truth and plausibility

As discussed in previous chapters, LLMs do not aim to discover or verify truths. Rather, they are designed to predict the most statistically probable sequence of words in response to a given prompt. This distinction has profound epistemological consequences. Unlike scientific inquiry or journalistic investigation, which are structured around truth-seeking and evidence, LLMs are optimized for linguistic coherence. As a result, they often produce fluent, contextually appropriate responses that may nevertheless lack factual accuracy.

1.1 The epistemic instability of AI-generated content

One of the most pressing concerns arising from this is the problem of faithfulness—the degree to which LLM outputs correspond to reality. Because these models are not explicitly trained to track truth conditions, they can generate text that is plausible yet ungrounded (see Chapter 1). This dynamic recalls Harry Frankfurt's (2005) notion of bullshit—discourse that is indifferent to truth and is aimed not necessarily at deception but at persuasion through appearance. Unlike lies, which require knowledge of the truth in order to distort it, bullshit operates independently of factuality. Similarly, while LLMs can sometimes distinguish between accurate and inaccurate statements, their primary optimization is for coherence, which means they may produce outputs that sound authoritative but remain untethered from external reality. This artificial fluency makes AI-generated misinformation particularly insidious: by default, it emerges not from deliberate ideological intent but from the statistical structure of language itself—though it can, of course, be guided by more conscious ideological goals as well.

This has far-reaching implications for epistemic authority and the institutions traditionally responsible for knowledge validation. In predigital environments, trust in information was mediated by various gatekeepers—scientific bodies, academic publishers, editorial boards, journalistic standards. While these structures were often biased or flawed, they provided mechanisms for verification, accountability, and critical scrutiny. LLMs, by contrast, circumvent these hierarchies. Their outputs, indistinguishable from human writing, erode the boundary between expertise and lin-guistic plausibility. As Jean-François Lyotard (1984) observes in *The Postmodern Condition*, the collapse of grand narratives and authoritative discourses leads to an epistemic landscape charac-terized by fragmentation and performativity. LLMs accelerate this collapse by generating authoritative-seeming text without under-going any epistemic labor—no research, no dialogue, no contesta-tion. In such a context, the line between knowledge and mimicry becomes increasingly blurred.

These strong dynamics are further amplified by the biases embedded in training data and the fragmented nature of online discourse. LLMs are trained on massive datasets that inevitably reflect the biases, omissions, and distortions of human com-munication (see Chapter 7). When these inputs include low-quality, polarized, or misleading sources, the models reproduce and amplify those distortions. This recursive loop mirrors Cass Sunstein's (2001) concept of *cyberbalkanization*, where online platforms reinforce ideological silos by filtering out dissenting viewpoints. While many users consume news from diverse out-lets, preference algorithms and recommendation systems tend to consolidate existing worldviews. With LLMs, this effect is intensi-fied: Not only do they reflect existing patterns but they generate new content based on those same patterns, reinforcing narrative bubbles and making misinformation self-perpetuating.

This detachment from human epistemic norms presents a para-dox. Although LLMs are trained on human-generated language, their mode of operation diverges fundamentally from human rea-soning. They do not engage in processes of verification, debate, or evidentiary scrutiny. Instead, they simulate the outputs of such processes without performing the underlying cognitive work. As AI-generated content becomes more prevalent in everyday com-munication, education, journalism, and public discourse, a critical question arises: How will human societies adapt their epistemic frameworks to accommodate these new nonhuman knowledge agents? If credibility is increasingly determined by linguistic flu-ency rather than by evidence or justification, there is a growing risk that truth itself will be subordinated to coherence.

This concern echoes Michel Foucault's (1981) analysis in *The Order of Discourse*, where he argues that knowledge is always mediated by systems of power that govern what counts as legitimate discourse. In an era dominated by predictive algorithms, the authority of human reasoning may be displaced by computational logics that prioritize pattern recognition over epistemic rigor. When LLMs generate immense volumes of synthetic text, they do not merely add information to the public sphere—they participate in shaping what is thinkable and sayable, often without transparency or contestability.

1.2 Human judgment and the reception of AI-generated content

While LLMs do not themselves engage in epistemic labor, their outputs are interpreted and evaluated by human users—who bring their own heuristics, expectations, and vulnerabilities to the encounter. Studies in psychology and media studies show that people often conflate fluency and confidence with credibility, especially in online environments (Pennycook and Rand, 2019). This creates a dangerous cognitive shortcut: Content that *looks* legitimate is more likely to be *believed*, regardless of its factual accuracy.

Users tend to rely on superficial cues such as formatting, tone, or presence of citations to judge credibility. In many cases, AI-generated outputs replicate these features convincingly—sometimes better than human writers. As a result, misinformation produced by LLMs may be adopted not because users trust the system explicitly but because it aligns with preexisting schemas for what 'reliable information' looks and sounds like.

Furthermore, the impact of such content is shaped by interpretive communities—social groups, online networks, and ideological clusters that influence how messages are received. Marwick and Lewis (2017), in their research on online radicalization, show how misinformation does not operate in a vacuum: It is embedded in cultural practices, narratives, and social trust structures that determine whether a claim is accepted, rejected, or amplified.

In this light, the epistemological threat of LLMs is not merely their capacity to generate misinformation, but their ability to colonize the form of knowledge—to produce outputs that pass as credible, authoritative, and neutral, while bypassing the institutional and cognitive scaffolding that once supported those attributes.

More profoundly, the epistemological challenge posed by LLMs is not reducible to a binary of true versus false. It concerns the deeper processes by which truth is constructed, validated, and disseminated. As human and machine-generated content become increasingly indistinguishable, we must reevaluate the criteria by which knowledge is assessed. If traditional gatekeeping mechanisms falter and epistemic authority becomes diffused or opaque, new frameworks are needed—ones that balance technical tools (such as fact-checking systems) with philosophical reflection on the nature of digital truth. Navigating these challenges will require a renewed emphasis on critical reasoning, media literacy, and transparent algorithmic design. But, more fundamentally, it requires recognizing that in an age of synthetic fluency, credibility is something that must be established, not taken for granted. The future of knowledge depends not only on what machines can generate but also on how humans choose to interpret, contest, and trust what they read.

2 Misinformation, disinformation, and the automation of narrative control

The rise of generative AI presents an epistemological crisis that goes beyond isolated falsehoods by automating the production of persuasive narratives and transforming how truth circulates in digital environments. It significantly alters how false narratives emerge, propagate, and influence public perception, blurring the line between misinformation—unintentional inaccuracies caused by human or algorithmic error—and disinformation, which involves deliberate efforts to mislead or manipulate for strategic ends (Wardle and Derakhshan, 2017). With the unprecedented speed and scale at which generative AI can produce fluent and convincing text, distinguishing honest mistakes from intentional deception becomes increasingly difficult. This section examines how generative models contribute to both misinformation and disinformation, and how they reshape the broader conditions under which narrative authority is produced and received.

2.1 Automated propaganda and political manipulation

AI-driven technologies provide unprecedented capabilities for automated propaganda and large-scale political manipulation. Unlike traditional forms of propaganda, which rely heavily on centralized, human-driven campaigns, AI facilitates decentralized, automated content production that can rapidly adapt to changing narratives or audience reactions. Automated bots, generative language models, and deepfake technologies have already been weaponized in electoral contexts and geopolitical conflicts to influence public opinion, destabilize political institutions, and erode democratic trust (Bradshaw and Howard, 2019). These technological advances permit malicious actors to orchestrate complex influence campaigns with a high degree of precision, making it increasingly difficult for users, platforms, and regulatory bodies to detect and counteract disinformation effectively (Goldstein et al., 2023). As discussed in the introduction, the war in Ukraine has provided a stark illustration of these dynamics, with large-scale propaganda being used to spread misleading narratives about the conflict, shaping public perceptions through social media amplification and automated content distribution.

Jacques Ellul (1965, 1973) describes modern propaganda as a systematic method for manipulating perceptions by continuously immersing individuals in constructed narratives. Generative AI drastically intensifies Ellul's notion of propaganda by enabling algorithmically driven narrative generation, capable of producing a continuous stream of persuasive yet deceptive content tailored to individual biases and emotional triggers. This transformation is particularly evident in the emergence of automated feedback loops, where AI-generated disinformation is not only spread through troll farms and bot networks but also reabsorbed by web crawlers that feed LLMs, subtly altering their responses to key political and historical questions.

For example, as said in the introduction, recent investigations have revealed that the Pravda network, a coordinated Russian propaganda operation consisting of over 150 interconnected websites, has deliberately flooded the internet with fabricated news articles in an effort to manipulate public discourse and influence AI systems. Producing millions of articles annually, the network targets web crawlers used to train LLMs, a tactic now referred to as 'LLM grooming.' This disinformation is then inadvertently incorporated into AI training data, allowing false or misleading narratives to resurface in chatbot-generated responses. A 2024 study by NewsGuard found that major AI chatbots—including ChatGPT-4, Google's Gemini, and Microsoft Copilot—reproduced disinformation sourced from the Pravda network in more than 30% of relevant prompts.[1] This process blurs the distinction between truth and falsehood at the level of both content and credibility, embedding propaganda within AI-driven knowledge systems and complicating efforts to maintain informational integrity.

Beyond the geopolitical context, AI-driven propaganda techniques have also been deployed in domestic political arenas. For example, NewsGuard reported that during Ghana's 2024 general election, a network of 171 fake accounts—likely created using ChatGPT—disseminated AI-generated content to influence voters and discredit opposition figures. Similar tactics have been observed in other democracies, where AI-enhanced disinformation campaigns have targeted elections, referenda, and public health crises, exploiting algorithmic recommendation systems to maximize engagement and virality. These developments highlight

---

[1] https://www.newsguardtech.com/special-reports/generative-ai-models-mimic-russian-disinformation-cite-fake-news.

the growing challenge posed by AI-powered propaganda: It not only intensifies the scale and reach of influence operations but also erodes public trust in information ecosystems, creating an environment where distinguishing between authentic and manipulated content becomes increasingly difficult.

2.2 AI credibility, persuasion, and user agency

The epistemic impact of AI-generated misinformation cannot be fully understood without considering how users interpret, respond to, and interact with such content. The issue is not solely what LLMs produce but how these outputs are received, shared, and contested by individuals and communities. Studies in science and technology studies (STS) emphasize the coconstruction of knowledge between technologies and their users: Credibility is not an inherent property of information but something negotiated in context (Jasanoff, 2004). This means that AI-generated content does not exert influence independently—it becomes epistemically potent through interpretation, circulation, and social uptake.

In the realm of misinformation, cognitive heuristics and affective responses play a critical role in shaping how users assess credibility. One of the most studied mechanisms is the illusory truth effect, which demonstrates that repeated exposure to a claim increases its perceived accuracy, even when it is demonstrably false (Fazio et al., 2015). In highly networked environments, AI-generated misinformation can be reshared, quoted, reworded, and echoed across multiple platforms, gaining epistemic weight through repetition. Importantly, the persuasiveness of such content often depends less on factual accuracy than on stylistic familiarity, emotional resonance, or ideological alignment.

Furthermore, many users operate within informational ecosystems shaped by algorithmic curation and social affinity. In such echo chambers, LLM outputs that reinforce existing beliefs may be interpreted not as synthetic responses but as confirmation of what is already presumed to be true. Research by Marwick and

Lewis (2017) on disinformation and online radicalization high-lights how misinformation, especially when tailored to resonate with identity or group loyalty, is often adopted not because it is convincing on evidentiary grounds but because it validates the user's worldview. In these contexts, what prevails is not always the factual status of the information, but the sense of belonging it fosters—where sharing (pseudo)information becomes a way of affirming allegiance to a community defined as much by shared values as by opposition to dominant narratives (boyd, 2017).

During the COVID-19 pandemic, social media platforms saw a surge in misleading narratives about vaccine safety, many of which resonated with wellness and antiestablishment communi-ties. While there is no confirmed case of AI-powered bots generat-ing such content on Facebook specifically, studies have shown that generative models can produce plausible antivaccine misinforma-tion when prompted.[2] Research also highlights that users often found such content credible not because of scientific grounding but because it mirrored the rhetorical style and values of their communities (Memon and Carley, 2020). In such cases, the per-suasive power of misinformation stems less from factual claims and more from stylistic familiarity and cultural alignment—a dynamic that AI can easily amplify.

Crucially, however, not all users are passive recipients of AI-generated content. Interpretive agency remains a vital part of the equation. In many cases, resistance to misinformation emerges through community-driven fact-checking efforts, educational interventions, and platform-level flagging or downranking of sus-picious content. Initiatives like Wikipedia's editorship model, or the work of digital literacy NGOs, exemplify how collective epis-temic practices can function as counterweights to the rapid spread of computationally generated disinformation.

Yet these efforts are far from evenly distributed. Access to media literacy resources, critical reasoning education, and trusted

---

[2] https://apnews.com/article/technology-science-business-artificial
-intelligence-afb4618ff593db9e3e51ecbd91dc3eef.

institutional guidance is often shaped by socioeconomic, cultural, and political inequalities. In contexts where trust in institutions is already low—or where information infrastructures are weak— misinformation is more likely to gain traction. As such, understanding how different communities interact with AI-generated text is essential not only for diagnosing the broader impact of LLMs but for designing more equitable and effective responses. This means going beyond technical solutions to consider cultural attitudes toward expertise, authority, and truth.

Ultimately, the epistemological stakes of LLMs lie not only in their ability to generate synthetic discourse but in how that discourse is read, used, and contested by human actors. Any attempt to mitigate misinformation must attend to these interpretive dynamics and recognize that meaning is not imposed by technology alone but coproduced in the human–machine interface.

2.3 The speed and scale of AI-driven misinformation diffusion

Historically, misinformation diffusion was constrained by the limitations inherent in human communication and coordination. In contrast, generative AI systems automate and exponentially accelerate misinformation spread, rapidly overwhelming traditional fact-checking and verification mechanisms. This phenomenon exemplifies what Manuel Castells (1996) describes as the network society, in which digital connectivity enables instantaneous and globalized flows of information. With AI-generated misinformation, however, these flows transform into torrents of automated falsehoods, capable of reaching millions within seconds and saturating the information environment to the point of informational, and ultimately epistemic, overload.

A striking example occurred during the 2024 US presidential election, when an AI-generated robocall mimicking President Joe Biden's voice was circulated in New Hampshire, falsely urging Democratic voters to 'save their vote' by staying home during the primary. The deepfake audio, produced using voice-cloning technology, was traced back to a political consultant who was later fined for election interference. Despite swift responses from election officials and fact-checkers, the message had already reached a significant portion of voters before corrective mechanisms could be deployed. This incident underscores how AI-generated misinformation can exploit the speed and scale of digital communication, outpacing traditional fact-checking and oversight systems. As Ferrara (2024) emphasizes, such rapid dissemination fosters environments in which misinformation circulates widely before verification efforts can intervene, thereby threatening the integrity of public discourse and democratic participation.

Moreover, the widespread adoption of AI-driven chatbots exacerbates the problem. During the same election cycle, automated disinformation campaigns deployed LLM-powered bots to engage with users on forums, reinforcing false narratives in a manner indistinguishable from genuine human discourse. This tactic effectively bypasses traditional epistemic gatekeepers, as misinformation is no longer constrained to overtly fake content but is embedded within everyday online conversations, further amplifying societal vulnerability to systemic manipulation.

This dynamic points to the urgent need for proactive mitigation strategies, such as real-time AI detection systems, improved regulatory frameworks, and public education initiatives that account for the evolving nature of misinformation in the age of generative AI.

2.4 AI, hyperreality, and the posttruth condition

Generative AI technologies also contribute significantly to the intensification of what Baudrillard (1994) terms hyperreality—a condition in which representations no longer correspond to an objective reality but instead constitute their own autonomous reality. Baudrillard argues that hyperreal narratives possess a greater potency and persuasive influence precisely because they do not rely on external reference points; instead, they achieve authority through their sheer plausibility and internal coherence. AI-generated misinformation exemplifies this hyperreality by producing texts that appear not merely believable but epistemically authoritative, despite potentially having no factual grounding whatsoever (Fuller, 2018).

Baudrillard famously illustrates hyperreality through the example of Disneyland, which, in his analysis, functions not simply as an amusement park but as a simulation that conceals the fact that the 'real' America itself operates like a vast Disneyland (Baudrillard, 1994). In this sense, Disneyland is not a counterpoint to reality but a paradigm of how the boundaries between the simulated and the real collapse. More recently, cultural phenomena like *Squid Game* demonstrate similar dynamics. The intense global engagement with the fictional world of *Squid Game*—from fan theories about character arcs to speculation about potential continuations—blurs the line between entertainment and reality. Audiences engage with the series not merely as fiction but as an immersive universe with real social implications, reinforcing hyperreal environments where mediated representations feel as tangible and consequential as factual events.

This tendency is not confined to entertainment. Political discourse has increasingly adopted hyperreal characteristics, as seen in figures like Donald Trump, who frequently invoked the term 'fake news' to delegitimize verifiable information while promoting alternate narratives detached from empirical evidence. Claims such as Ukraine attacking Russia (contrary to documented facts) and the denial of anthropogenic climate change exemplify deliberate manipulations of reality to fit ideological ends. These distortions create self-sustaining informational ecosystems where internal consistency supersedes factual accuracy—a phenomenon Baudrillard's theory helps to illuminate. Another illustration of this dynamic occurred during Hurricane Milton in October 2024, when Trump and his supporters circulated false claims accusing the Biden administration of neglecting the crisis, despite well-documented emergency responses. This disinformation campaign, amplified by AI-generated imagery, further blurred the boundaries between factual events and fabricated narratives, reinforcing a hyperreal environment where representation overrides reality.

It is essential to recognize that, while generative AI and LLMs are adept at producing hyperreal content, they are not the sole architects of hyperreality. Rather, such technologies amplify existing tendencies within media and political culture. Their outputs flourish within environments already predisposed to disinformation and detachment from factual grounding. Hannah Arendt (1967) warned of the dangers inherent in political cultures untethered from factual reality, observing that totalitarian regimes thrive in contexts where truth becomes subordinate to ideological narratives. Arendt emphasized that, when facts lose their status as common ground, the public sphere becomes vulnerable to manipulation and control. In this light, generative AI operates less as the originator of hyperreality and more as a catalyst that accelerates its proliferation within sociopolitical contexts already marked by epistemic instability.

2.5 Toward a new epistemological framework for digital truth claims

Given these profound shifts, society faces the urgent challenge of reassessing how truth claims are verified, contested, and validated in the context of automated content generation. The erosion of epistemic clarity demands more than technical countermeasures such as algorithmic moderation, fact-checking technologies, or regulatory interventions—though these remain necessary. It also requires a philosophical reevaluation of truth, knowledge, and trust. As Bruno Latour (2004) suggests, the stability of facts depends heavily on networks of trust, institutional consensus, and societal agreement. Generative AI disrupts these networks by producing knowledge claims independently of traditional epistemic scrutiny or institutional vetting, thus fundamentally destabilizing established norms of factual verification.

In this context, responding to the challenges of misinformation and disinformation requires more than reinforcing old paradigms.

It calls for the development of what might be termed epistemic dispositions or normative practices—that is, the habits, attitudes, and interpretive skills that help individuals and communities engage critically with knowledge claims in increasingly automated and uncertain environments. These include commitments to transparency, algorithmic accountability, critical media literacy, and informed skepticism toward computationally generated content.

Scholars such as Luciano Floridi (2019) have emphasized the importance of cultivating these dispositions not only at the individual level but across institutions, technologies, and educational systems. Rather than assuming the epistemic integrity of digital outputs, users must be equipped to question how knowledge is produced, whose interests it serves, and by what standards it ought to be judged. Developing these capacities will require an interdisciplinary effort—combining philosophical reflection, technological design, and political engagement—to foster resilient and reflective epistemic communities capable of navigating, and critically interrogating, the evolving dynamics of the digital information ecosystem, even though we have seen that enhanced digital literacy is far from sufficient when facing conspiracy thinking and the effects of group identity.

This exploration thus serves as a foundation for the following section, which will examine the deeper sociopolitical implications of these epistemic transformations, focusing specifically on how generative AI affects public trust, epistemic authority, and democratic accountability. In a world where generative AI reshapes how knowledge appears, circulates, and gains legitimacy, epistemic vigilance must be reimagined not as a matter of passive reception but of active, collective reconstruction.

3 The role of platform governance and policy

The ethical and political stakes of AI-generated misinformation suggest the need for regulatory and governance responses. Yet, as with all attempts to regulate speech, the balance between moderation, freedom of expression, and accountability remains precarious. The proliferation of AI-generated text, with its capacity for mass production and seamless integration into human discourse, complicates efforts to control its impact. Moreover, information governance is inherently contested: States, political leaders, and regulatory bodies are not neutral actors but are themselves subject to disputes over legitimacy, authority, and influence. This makes the task of regulating information flows particularly challenging.

3.1 Artificial discourse and the erosion of epistemic trust

Traditional fact-checking mechanisms, though valuable, face considerable difficulties in keeping pace with the scale and velocity of AI-generated content. The logic of misinformation has shifted: No longer a matter of isolated falsehoods to be debunked, it has become a vast, automated churn of synthetic narratives that mimic human reasoning. Hannah Arendt warned that the erosion of a shared reality is a prelude to political instability—today, the challenge is not just the falsification of facts but the saturation of discourse with plausible fabrications.

A more sustainable approach may lie in fostering AI literacy. If users can critically evaluate sources and recognize patterns indicative of AI-generated misinformation, the impact of false narratives may be mitigated at the interpretative level. Just as the Enlightenment philosophers championed education as a means of liberating individuals from dogma, digital literacy in the AI age could empower citizens to navigate an increasingly synthetic information landscape. However, literacy alone does not resolve the broader tensions of trust and authority—who determines what counts as credible information, and whose interests are served in the process?

Legal and regulatory responses to AI-generated misinformation have been proposed, but implementation remains fraught with complexity. Watermarking AI-generated texts (see Chapter 3),

transparency requirements for AI systems, and platform accountability measures all offer potential safeguards. However, information
flows are now global, decentralized, and often anonymized,
making enforcement difficult. Additionally, states and regulatory bodies do not operate in a vacuum—many are themselves
embroiled in political contestation, making their authority to regulate information a matter of ongoing debate.

Michel Foucault's notion of power-knowledge (1977, 1978) is
particularly relevant here: Governance over AI-generated content
is not merely a question of legal restriction but of who holds the
authority to determine truth. If platforms become the arbiters of
what is considered reliable, does this not create a new epistemic
monopoly? Conversely, if regulation is too loose, do we risk a
landscape where misinformation dominates, undermining public
trust in knowledge itself? Furthermore, states, especially in pluralistic societies, do not always have uniform control over information; different political factions may compete to impose their
own standards of truth, further complicating regulatory efforts.

The increasing presence of AI-generated content in journalism
and academic publishing raises profound questions about authorship, credibility, and the integrity of knowledge production. If
an AI-generated article makes it into a respected journal, does it
carry the same epistemic weight as one written by a human expert?
What happens when misinformation, generated by AI, is seam-
lessly woven into sources traditionally considered authoritative?

Here, Walter Benjamin's concern about the loss of the 'aura' of
originality in the age of mechanical reproduction remains relevant
(Benjamin, 2008 [1936]; see also Chapter 3). When AI not only
assists but autonomously generates text, the distinction between
an 'authentic' source and a fabricated one begins to blur. The problem is not just one of deception but of epistemology: How do we
preserve trust in knowledge when its production becomes indistinguishable from its manipulation? This issue becomes even more
pressing in a political environment where even human-generated
knowledge is often disputed, further underscoring the fragility of
any attempt to establish definitive informational authority.

The governance of AI-generated misinformation is not simply a technical challenge but a political struggle over knowledge and narrative authority in digital societies. The intersection of regulation, platform responsibility, and user literacy will influence the future of information ecosystems, but none of these measures can in isolation offer a complete solution. Moreover, because states and political leaders are themselves often the subject of contestation, their ability to control information is inherently limited. This limitation is not only structural but also desirable, since information should remain free, reflect the diversity of opinions, and not be subordinated to state control. Whether the future of AI-driven discourse fosters informed engagement or descends into an endless play of simulation remains an open question—one that is as much about philosophy as it is about policy.

3.2 Reclaiming human judgment in an AI-mediated world

The widespread use of AI in content generation compels us to reconsider fundamental philosophical and ethical questions about truth, authority, and human judgment in contemporary digital societies. As LLMs and other AI-driven systems increasingly mediate our access to information, the nature of knowledge itself comes into question. As we saw in the first part of this book, classical epistemology, which defines knowledge as justified true belief, would suggest that AI lacks the necessary grounding to produce knowledge.

A related concern is the automation of truth-making. AI's role in news production, scholarly publishing, and historical documentation raises questions about how truth is constructed and maintained in an era where machine-generated content is ubiquitous. If discourse is a form of power, as Michel Foucault (1981) famously argues, then AI's increasing influence over language suggests a new form of epistemic authority (here again, see the first part of the book, esp. Chapter 3). By shaping narratives, filtering information, and amplifying particular viewpoints, AI systems do not merely reflect reality but actively participate in its construction. The extent to which AI-driven systems reinforce existing biases, privilege certain perspectives, or subtly reshape public discourse remains an open question, but it is one that demands careful scrutiny.

In response to these challenges, some scholars and ethicists advocate for a renewed emphasis on human judgment, critical thinking, and epistemic responsibility. AI should not be treated as an autonomous arbiter of truth but as a tool embedded within broader deliberative and contestable knowledge systems. This calls for mechanisms that enhance transparency, allow for contestation, and maintain space for human verification and interpretative agency. Rather than ceding authority to automated systems, we might envision AI as a supplement to human inquiry—one that can assist but not replace the complex, value-laden processes through which truth is assessed and affirmed.

The ethical stakes of these developments are significant. If knowledge is increasingly mediated by AI, there is a risk that epistemic authority will become more opaque and concentrated in the hands of those who control these technologies. Ensuring that AI remains accountable, interpretable, and aligned with democratic values will require a concerted effort across disciplines, from philosophy and ethics to computer science and policymaking. The challenge is not simply to refine AI's technical capabilities but to reflect on the broader implications of its integration into human epistemic practices. In doing so, we might reclaim a vision of knowledge production that prioritizes transparency, critical engagement, and the indispensable role of human judgment in an AI-mediated world.

3.3 The overabundance of AI ethics charters and their limited impact

Over the past decade, a proliferation of charters, good practice guidelines, and white papers has sought to define ethical standards for AI, particularly LLMs. Institutions ranging from international organizations (such as UNESCO and the European Union) to private companies and research centers have published extensive documents outlining principles such as transparency, fairness, accountability, and human oversight. These initiatives aim to provide a normative framework that ensures AI technologies are developed and deployed responsibly, mitigating potential harms while maximizing benefits.

Among the most widely cited documents is the Ethics Guidelines for Trustworthy AI (European Commission, 2019), which outlines seven key requirements for AI systems, including human agency, technical robustness, and accountability. UNESCO's Recommendation on the Ethics of Artificial Intelligence (2021) similarly emphasizes inclusivity, environmental sustainability, and respect for cultural and linguistic diversity. Industry actors have also contributed, with OpenAI's policy statements on responsible AI use[3] and Google's AI Principles,[4] which emphasize fairness and explainability. These efforts highlight a shared commitment to ethical AI development across public and private sectors.

In recent years, however, the limitations of nonbinding ethical frameworks have become more apparent (Hagendorff, 2020), prompting a shift toward enforceable legal instruments. Most notably, the European Union has taken a leading role with the adoption of the AI Act,[5] approved in 2024 and set to become the first comprehensive, legally binding regulatory framework for AI globally. The act introduces a risk-based classification system for AI applications, imposes strict obligations for high-risk systems—including many LLMs—and mandates transparency requirements for general-purpose AI. It represents a move from voluntary ethics to hard law, backed by compliance mechanisms and penalties. Complementary efforts such as the EU's AI Liability Directive and

---

[3] https://openai.com/safety.
[4] https://ai.google/responsibility/principles.
[5] https://artificialintelligenceact.eu.

Digital Services Act[6] further reflect an evolving regulatory ecosystem that links ethical principles to enforceable standards.

Despite these advancements, challenges remain. A key issue is the continued proliferation of overlapping and sometimes contradictory frameworks—many of which are still produced by industry actors or regional consortia without coordination. As a result, practitioners often face uncertainty about which standards to adopt and how to operationalize them. Moreover, even legally binding frameworks such as the AI Act face criticisms: For instance, some observers note that enforcement capabilities across EU member states remain uneven, and loopholes may persist for powerful corporate actors. The risk of ethics-washing—the strategic use of ethical language for reputational gain rather than substantive governance—remains present (Bietti, 2020), particularly outside the scope of formal regulation.

Contextual sensitivity also remains a challenge. What constitutes responsible AI use in health care or public services may differ significantly from what is required in creative, educational, or low-stakes consumer applications. Yet, many guidelines still adopt a broad, universalist tone that risks flattening these differences. Earlier initiatives, such as the Montreal Declaration for Responsible AI (2018), offered a compelling ethical vision but lacked implementation mechanisms—highlighting the gap between principles and policy.

Nonetheless, these documents have played an important discursive role in shaping public debate and laying the groundwork for regulatory innovation. The transition from voluntary charters to binding frameworks—especially in the European context—marks a turning point in global AI governance. Moving forward, the challenge is to ensure that these instruments are not only enforceable but also adaptable to rapidly evolving technological, cultural, and political contexts. This includes developing sector-specific compliance strategies, robust auditing mechanisms, and international

---

6 https://digital-strategy.ec.europa.eu/en/policies/digital-services-act -package.

coordination efforts that extend beyond Western-centric paradigms. Only then can ethical aspirations be meaningfully translated
into institutional practice.

4 AI governance: competing interests and unresolved challenges

The governance of AI is increasingly recognized as a fundamental
issue, yet consensus on how to regulate and structure AI development remains elusive. The principles guiding AI governance often
include transparency, accountability, fairness, and human oversight,
reflecting ethical concerns that have long been debated in philosophy and political theory. Classical notions of governance from theorists such as Foucault and Habermas provide critical insights into
the ways power, discourse, and regulation shape the control of technological systems. Meanwhile, governance studies have drawn on
international regulatory frameworks to examine how institutions
manage technological risks and opportunities, highlighting both
the necessity and the difficulty of enforcing global AI policies.

Critical studies of AI have pointed to the asymmetry of power
in governance debates, where dominant technological actors—
whether multinational corporations or state entities—frame the
discourse around safety, innovation, and ethical use. Scholars
in critical AI studies, such as Kate Crawford (2021) and Ruha
Benjamin (2019), emphasize that governance structures are not
neutral but rather reinforce existing sociopolitical hierarchies.
This is in line with broader critiques in STS, where governance
is seen as a battleground of competing economic, political, and
ideological forces rather than a purely technical or ethical issue.
The governance of AI, therefore, cannot be detached from questions of power, labor, and geopolitical influence.

In practice, AI governance remains fragmented, with different
nations and regional blocs seeking to assert their own models and
priorities. The European Union's AI Act, for instance, attempts to
establish strict regulations emphasizing fundamental rights and

consumer protection, while China has opted for a more central-ized approach that aligns AI regulation with state security and economic ambitions. The United States, by contrast, has favored a market-driven approach, with a preference for industry self-regulation and targeted interventions rather than comprehensive federal oversight. These divergent strategies reflect deeper politi-cal and economic interests, with each entity attempting to position itself as a leader in the AI domain.

This multiplicity of governance models reveals the absence of a unified global framework, raising concerns about regulatory arbitrage and uneven ethical standards. The challenge is exacer-bated by the transnational nature of AI systems, which operate across borders in ways that defy traditional regulatory mecha-nisms. While initiatives such as the OECD AI Principles and UN-driven discussions aim to foster international cooperation, their effectiveness remains limited in the face of nationalistic strategies and corporate lobbying. Recent high-profile events, such as the 2023 AI Safety Summit in London and the 2025 Paris AI Govern-ance Forum, have produced aspirational declarations but failed to deliver binding commitments. Despite rhetorical agreement on principles like transparency and safety, each participating coun-try has largely defended its own strategic and economic interests, undermining prospects for cohesive global regulation.

Fundamentally, AI governance is not just a matter of defin-ing ethical principles or designing regulatory mechanisms; it is a deeply political issue that reflects broader struggles over eco-nomic dominance, security, and societal values. The ongoing debates in governance studies and critical AI scholarship high-light the need for approaches that move beyond idealistic ethical charters to concrete policies that account for power dynamics and geopolitical realities.

References

Arendt, H. (1967). Truth and politics. *The New Yorker*, February 25, 1967.

Baudrillard, J. (1994). *Simulacra and simulation* (Trans. S. F. Glaser). University of Michigan Press.

Benjamin, W. (2008 [1936]). The work of art in the age of its technological reproducibility (E. F. N. Jephcott and H. Zohn, Trans.). In H. Eiland and M. W. Jennings (Eds.), *The work of art in the age of its technological reproducibility, and other writings on media* (pp. 19–55). Harvard University Press.

Benjamin, R. (2019). *Race after technology: Abolitionist tools for the New Jim Code*. Polity Press.

Bietti, E. (2020). From ethics washing to ethics bashing: A view on tech ethics from within moral philosophy. In *Proceedings of the 2020 Conference on Fairness, Accountability, and Transparency (FAT* '20), New York, NY, USA* (pp. 210–219). Association for Computing Machinery. https://doi.org/10.1145/3351095.3372860

boyd, d. (2017). You think you want media literacy… do you? *Data & Society*. Republished on Medium. https://medium.com/datasociety-points/you-think-you-want-media-literacy-do-you-7cad6af18ec2

Bradshaw, S. and Howard, P. N. (2019). The global disinformation order: 2019 global inventory of organised social media manipulation. *Computational Propaganda Project by the Oxford Internet Institute, University of Oxford (CyberTroop Report 19)*. https://demtech.oii.ox.ac.uk/wp-content/uploads/sites/93/2019/09/CyberTroop-Report19.pdf

Castells, M. (1996). *The rise of the network society* (Vol. 1). Blackwell.

Crawford, K. (2021). *Atlas of AI: Power, politics, and the planetary costs of artificial intelligence*. Yale University Press.

Ellul, J. (1965). *The technological society* (J. Wilkinson, Trans.). Vintage.

Ellul, J. (1973). *Propaganda: The formation of men's attitudes* (K. Kellen and J. Lerner, Trans.). Vintage.

European Commission (2019). *Ethics guidelines for trustworthy AI*. Publications Office of the European Union. https://digital-strategy.ec.europa.eu/en/library/ethics-guidelines-trustworthy-ai

Fazio, L. K., Brashier, N. M., Payne, B. K., and Marsh, E. J. (2015). Knowledge does not protect against illusory truth. *Journal of Experimental Psychology: General, 144*(5), 993–1002. https://doi.org/10.1037/xge0000098

Ferrara, E. (2024). *Charting the landscape of nefarious uses of generative artificial intelligence for online election interference*. https://ssrn.com/abstract=4883403

Floridi, L. (2014). *The fourth revolution: How the infosphere is reshaping human reality*. Oxford University Press.

Floridi, L. (2019). Establishing the rules for building trustworthy AI. *Nature Machine Intelligence, 1*, 261–262. https://doi.org/10.1038/s42256-019-0055-y

Foucault, M. (1977). *Discipline and punish: The birth of the prison* (A. Sheridan, Trans.). Pantheon Books.

Foucault, M. (1978). *The history of sexuality, Volume 1: An introduction* (R. Hurley, Trans.). Pantheon Books.

Foucault, M. (1981). *The order of discourse* (R. Young, Trans.). In R. Young (Ed.), *Untying the text: A post-structuralist reader* (pp. 48–78). Routledge.

Frankfurt, H. G. (2005). *On bullshit*. Princeton University Press.

Fuller, S. (2018). *Post-truth: Knowledge as a power game*. Anthem Press.

Goldstein, J. A., Sastry, G., Musser, M., DiResta, R., Gentzel, M., and Sedova, K. (2023). Generative language models and automated influence operations: Emerging threats and potential mitigations. arXiv preprint. https://arxiv.org/abs/2301.04246

Hagendorff, T. (2020). The ethics of AI ethics: An evaluation of guidelines. *Minds and Machines, 30*(1), 99–120. https://doi.org/10.1007/s11023-020-09517-8

Jasanoff, S. (Ed.). (2004). *States of knowledge: The co-production of science and the social order*. Routledge.

Latour, B. (2004). *Politics of nature*. Harvard University Press.

Lyotard, J.-F. (1984). *The postmodern condition: A report on knowledge* (G. Bennington and B. Massumi, Trans.). University of Minnesota Press.

Marwick, A., and Lewis, R. (2017). *Media manipulation and disinformation online*. Data & Society Research Institute. https://datasociety.net/library/media-manipulation-and-disinfo-online

Memon, S., and Carley, K. M. (2020). Characterizing COVID-19 misinformation communities Using a novel Twitter dataset. arXiv preprint. https://arxiv.org/abs/2008.00791

Pennycook, G., and Rand, D. G. (2019). Lazy, not biased: Susceptibility to partisan fake news is better explained by lack of reasoning than by motivated reasoning. *Cognition, 188*, 39–50. https://doi.org/10.1016/j.cognition.2018.06.011

Sunstein, C. R. (2001). *Republic.com*. Princeton University Press.

UNESCO. (2021). *Recommendation on the ethics of artificial intelligence.* United Nations Educational, Scientific and Cultural Organization. https://unesdoc.unesco.org/ark:/48223/pf0000381137

Université de Montréal. (2018). *Montreal declaration for a responsible development of artificial intelligence.* https://www.montrealdeclaration -responsibleai.com

Wardle, C., and Derakhshan, H. (2017). *Information disorder: Toward an interdisciplinary framework for research and policy making.* Council of Europe. https://rm.coe.int/information-disorder-report /1680764666

Zannettou, S., Sirivianos, M., Blackburn, J., and Kourtellis, N. (2019). The web of false information: Rumors, fake news, hoaxes, clickbait, and various other shenanigans. *Journal of Data and Information Quality, 11*(3), Article 10. https://doi.org/10.1145/3309699

CHAPTER 9

Ethics at scale

This final chapter departs somewhat from the preceding philosophical inquiries into language, cognition, bias, and truth. It turns toward the lived dimension of interacting with large language models (LLMs): how they affect us in practice, and what kind of future they are already shaping. Though the topics may seem heterogeneous—ranging from labor and responsibility to ecology and cultural rights—they are united by the ethical and political questions that emerge when language models become embedded in everyday life.

From research labs to classrooms, customer support chatbots to automated legal assistants, LLMs have moved rapidly from experimental tools to general-purpose infrastructures. They are now deployed in education, health care, law, journalism, creative industries, public administration, and even religious practice. What began as linguistic models have become cognitive companions, content filters, and decision aids—often without users being fully aware of the scope of their integration. As Suchman (2007) argues in a different context, technologies never merely *mediate* human activity—they also *configure* it. LLMs, by virtue of their

How to cite this book chapter:
Poibeau, T. 2025. *Understanding Conversational AI: Philosophy, Ethics and Social Impact of Large Language Models.* Pp. 203–227. London: Ubiquity Press. DOI: https://doi.org/10.5334/bde.j. License: CC BY-NC 4.0

linguistic capabilities, do not just assist with language; they shape how language is used, which forms of expression are deemed appropriate, and which become invisible.

The shift from tool to infrastructure is not merely semantic: It signals a transformation in scale and epistemic authority. Infrastructure, as Bowker and Star (1999) remind us, is what becomes invisible through habitual use—but remains deeply consequential. When LLMs become infrastructural, they embed certain normative assumptions—about grammar, politeness, relevance, or credibility—into the very fabric of communication. They do not just assist with writing or translation; they subtly govern what it means to be clear, correct, or persuasive. In this sense, the stakes of their design and deployment extend far beyond technical performance.

This chapter develops around a set of interwoven questions: What happens to human agency when language is increasingly mediated by machines? How do ethical sensibilities shift when moral reasoning is delegated to systems optimized for plausibility rather than understanding? What forms of representation, extraction, and exclusion are encoded in the data these models consume—and what kinds of epistemic authority do they quietly assume? Beyond questions of function and performance, the chapter asks what it means to live in a world where language models operate not only as tools but as infrastructures—shaping attention, redistributing power, and reconfiguring the conditions of communication itself. Rather than offering a single argument, the chapter traces these concerns across multiple domains—labor, judgment, ecology, cultural rights, and political economy—seeking to understand not only what LLMs do but what they make possible, permissible, or invisible. The aim is not simply to assess their impact but to reflect on how they alter the space of human thought and action.

1 Automation and agency

One of the most consequential transformations brought about by LLMs is their role in the automation of linguistic and cognitive labor. From writing reports to answering legal or medical questions, these systems increasingly operate not only as tools but as agents that perform or simulate complex acts of reasoning. This raises pressing philosophical questions: When LLMs mediate thought and communication, what remains of human agency? And how should responsibility be distributed in environments where decisions are shared between human and artificial actors?

At first glance, one might turn to Kantian autonomy—the idea that to be free is to act not merely on impulse or external influence but according to principles one has rationally and reflectively chosen. From this perspective, the concern is not simply that LLMs replace human labor but that they subtly reshape the conditions under which we exercise judgment and recognize ourselves as the authors of our thoughts and actions. In an essay from 1784, Kant framed Enlightenment as the 'exit from self-incurred immaturity,' a process marked by the courage to use one's own reason (Kant, 1996 [1784]). The proliferation of LLMs does not constitute a return to immaturity in any straightforward sense but it may encourage a kind of intellectual delegation that erodes the habit of critical reflection. When users routinely defer to model-generated outputs—because they are fast, plausible, and confidently expressed—they may begin to treat these outputs less as starting points for deliberation and more as ready-made conclusions. This shift, from active reasoning to passive selection among prestructured options, risks relocating the locus of agency: from asking 'What do I think?' to asking 'Which of these sounds right?' While this does not negate autonomy altogether, it complicates the terrain on which it is exercised, raising important questions about the evolving relationship between reasoning, authority, and technological mediation.

Don Ihde's (1990) phenomenology of technological mediation is particularly helpful here. Ihde argues that technologies never merely extend human capabilities—they transform the structure of perception and action. In his terms, LLMs would be hermeneutic technologies: They interpret the world for us, offering preprocessed representations of knowledge, sentiment, or argument.

As such, they shape what appears salient, credible, or relevant. In doing so, they subtly recalibrate our epistemic agency: We no longer confront raw information but a curated, generated version of it. The danger is not simply inaccuracy but in a gradual shift of interpretive authority from the human to the machine.

Bruno Latour's notion of delegation (1992) similarly helps to articulate the redistribution of agency in hybrid systems. For Latour, technologies are not neutral tools but actors in networks of action—they embody scripts and intentions that influence behavior. In this sense, an LLM embedded in a workflow (e.g., email drafting, content moderation, or medical triage) participates in shaping outcomes even if it lacks intentionality. When a professional defers to a model's suggestion—perhaps because it 'sounds right' or aligns with institutional protocols—the model becomes a quasiagent, enrolled in the moral and epistemic logic of the task. The delegation of tasks to LLMs is therefore also a delegation of norms: of what counts as appropriate language, a plausible answer, or a sufficient justification.

This automation of judgment does not occur in a vacuum. In many sectors, it is driven by institutional imperatives—efficiency, scalability, cost reduction—that reward automation regardless of its philosophical implications. But, from a humanistic standpoint, it demands scrutiny. If language is not merely a medium but a form of life—as Wittgenstein would suggest (see Chapter 1)—then outsourcing linguistic activity to LLMs is not ethically neutral. It risks narrowing the space for human deliberation, ambiguity, and disagreement—qualities essential to democratic agency and moral growth.

2 Ethical drift and moral deskilling

As LLMs become integrated into everyday decision-making, they are not only automating tasks but also altering the ethical topography in which those tasks are embedded. In many domains—content moderation, hiring, medical triage, or even judicial support—language models are used to filter, prioritize, and sometimes make decisions that once required human deliberation. This growing reliance on AI systems introduces a form of what we might call ethical drift: a gradual displacement of moral responsibility, not through explicit delegation but through habituated deferral.

At the heart of this concern is the phenomenon of moral deskilling. Drawing on Hubert Dreyfus's (1992) critique of AI, we can understand moral expertise not as the application of fixed rules but as a cultivated capacity for situated and embodied judgment. Dreyfus argues that expert knowledge is rooted in context-sensitive responsiveness—something irreducible to formal codification. Similarly, moral judgment depends on attentiveness to ambiguity, nuance, and the singularity of others. When AI systems, such as LLMs, offer quick and plausible responses in ethically charged situations, they may discourage the kind of slow, reflective engagement that ethical life demands. In this way, the convenience of AI risks eroding the very capacities— empathy, discernment, and moral imagination—on which ethical practice depends.

This concern is not hypothetical. Past studies (Citron and Pasquale, 2014), especially in medical and legal contexts, have shown how decision-support systems can subtly shift professional norms, leading users to align their reasoning with algorithmic outputs even when those outputs are contested or incomplete (Chen et al., 2024; Choudhury and Chaudhry, 2024). This effect is amplified when the system's logic is opaque: Black-box models do not invite interpretation or dialogue but acceptance or rejection. The ethical danger is not just that humans might be replaced but that their capacity for judgment might be reshaped—narrowed by the framing, defaults, and authority of the system itself.

The tendency to offload moral labor onto technical systems reflects a broader cultural trend that Shannon Vallor (2016) describes as a crisis of moral attention. In a world saturated with digital automation, we risk becoming inattentive to the moral dimensions of our actions, treating ethical reflection as a resource-intensive luxury rather than a core human faculty. Vallor draws on Aristotelian virtue ethics to argue that flourishing in a technological age requires the cultivation of technomoral virtues: habits of mind and character that allow us to navigate technological complexity with responsibility and care.

What follows from this diagnosis is not a rejection of LLMs but a commitment to designing for ethical attentiveness. This involves keeping human beings meaningfully in the loop—not merely as overseers or fail-safes but as moral participants. It also calls for institutional structures that support ethical reflection: time, space, and norms that resist the logic of automation-as-default. As LLMs become entangled with social decision-making, the challenge is not just to govern them but to ensure that human users do not lose the moral competencies that make governance meaningful in the first place.

3 Data, consent, and cultural rights

The rise of LLMs has prompted renewed scrutiny of the relationship between data and power (see previous chapters), particularly in the context of linguistic, legal, and cultural representation. If LLMs are trained on the linguistic traces of billions of human interactions—books, websites, social media posts, oral transcripts— they inevitably raise pressing questions: Whose language is being modeled, under what conditions, and with what consequences (see Chapter 7)?

These patterns of exclusion are further intensified by the political economy of data. LLMs rely on vast corpora scraped from public and semipublic spaces, often without consent and with little regard for the contexts in which language is embedded.

Behind the apparent neutrality of 'training data' lie complex ethical, political, and increasingly legal concerns. Most large-scale language models are trained on vast, scraped datasets compiled without explicit consent from authors, communities, or data subjects. This includes copyrighted books, news articles, artworks, and oral traditions—often obtained without licensing or permission. While such practices are commonly justified under expansive interpretations of 'fair use' in the United States or 'text and data mining exemptions' in the European Union, these legal frameworks remain contested and incomplete. This process has even been described as 'data colonialism'—the appropriation of human expression and cultural production for computational value (Couldry and Mejias, 2019).

A prominent example is the lawsuit filed in the United States by major publishers—including the *New York Times*—against OpenAI and Microsoft, alleging unauthorized use of millions of copyrighted articles in the training of LLMs. The case highlights the potential for substantial financial liabilities, with damages potentially reaching into the billions. Such lawsuits are not merely about attribution or transparency—they challenge the very legality of current training practices and signal that the legitimacy of large-scale data scraping is now under serious judicial and regulatory scrutiny.

This emerging legal terrain has serious implications—not only for tech companies but for the entire creative ecosystem. When copyrighted material is reused without compensation, the economic basis of authorship is eroded. Writers, journalists, artists, and small publishers may find themselves in competition with synthetic outputs built on their own unpaid labor. The European Commission's proposed AI Code of Practice[1] encourages transparency and compliance with intellectual property law, but critics argue it leaves loopholes that may ultimately benefit large platforms over smaller rights-holders. Adding to these concerns, several major AI companies—such as Meta—have signaled they will not sign voluntary frameworks like the Code, framing them as burdensome or misaligned with innovation goals. This regulatory pushback highlights the widening gap between platform power and public accountability. The asymmetry is both financial and epistemic.

---

[1] https://digital-strategy.ec.europa.eu/en/policies/ai-code-practice.

Moreover, these practices disproportionately affect marginalized and Indigenous communities, whose cultural and linguistic
resources have historically been subject to extraction and mis-
representation. Many such communities have fought to protect
traditional knowledge from appropriation, distortion, or erasure—
whether by extractive academic research or exploitative digital
systems (see Liu [2024], for an overview in the case of LLMs).
When their languages are included in LLM training without consultation or benefit-sharing, the result is often not empowerment
but assimilation: Their epistemologies are flattened into mainstream linguistic patterns, stripped of context, and reintegrated
into systems they do not control.

This dynamic exemplifies what Miranda Fricker (2007) calls
epistemic injustice: a wrong done to someone specifically in their
capacity as a knower. In the case of LLMs, such injustice occurs
when communities are excluded from shaping how their linguistic and cultural practices are represented, interpreted, or reproduced. The harm is compounded when these same communities
are underrepresented in model outputs—reinforcing dominant
norms and marginalizing alternative forms of expression, storytelling, and reasoning.

But inclusion alone is not enough. Simply adding more 'diverse'
data to a training corpus does not resolve the deeper ethical issues
of consent, control, and cultural integrity. As scholars and activ-
ists in the data justice movement have argued (Taylor, 2017), what
matters is not just *what* data is included but *how, why,* and *by
whom.* This calls for a shift toward participatory approaches to
dataset curation and model design—ones that respect community
sovereignty, recognize the cultural specificity of language, and
enable opt-in rather than default inclusion.

Some initiatives have begun to address these challenges. Fully
open models developed by research groups such as the Allen
Institute for AI (e.g., OLMo[2]) and the TurkuNLP group (e.g.,

---

Poro[3]) have made notable efforts toward improving transparency. These projects go beyond simply releasing model weights—a common but limited form of openness—by also providing thorough documentation of training data sources, model architectures, and training procedures. This level of openness supports reproducibility and enables more meaningful scrutiny by the research community. However, broader structural changes are still needed. These may include enforceable consent mechanisms, ethical licensing frameworks, and financial compensation for rights-holders. While the recognition of collective cultural data rights—the idea that groups, not just individuals, have legal and moral claims over the use and distribution of their linguistic and cultural data—remains a compelling direction, it has yet to be meaningfully addressed even within many open model initiatives.

As regulatory frameworks evolve in Europe, North America, and elsewhere, the future of LLMs will increasingly be shaped not only by technical capacity but by legal accountability and cultural legitimacy. The central question is no longer just *what can be done* with data—but *what should be done*, in whose name, and to whose benefit.

4 Ecological and social externalities

While much of the ethical discourse around LLMs focuses on language, truth, or bias, there is growing recognition that these systems have *material* footprints—ecological, economic, and social—that extend far beyond the screen. The production, training, and deployment of large-scale models require enormous computational resources, specialized hardware, and a globalized labor force. While public narratives often celebrate the immateriality (or 'magic') of AI, there is growing awareness that these systems carry significant externalities. These material costs—long

---

[3] https://www.silo.ai/blog/poro-a-family-of-open-models-that-bring-european-languages-to-the-frontier.

obscured—are now increasingly recognized as central to any responsible evaluation of the societal role of LLMs.

4.1 The environmental cost of large language models

On the ecological front, the environmental cost of developing and deploying LLMs has become an urgent area of concern (Strubell et al., 2019). Training models such as GPT-3 and GPT-4 involves the consumption of massive computational resources, with corresponding impacts on energy use and carbon emissions. Estimates suggest that training GPT-3 alone required approximately 1,287 megawatt hours of electricity and produced over 550 metric tons of CO_2 emissions—comparable to the total emissions of an average American household over several decades (Patterson et al., 2021). As newer models scale into the hundreds of billions or even trillions of parameters, their resource demands grow exponentially, especially when models are retrained or fine-tuned for new tasks, languages, or user-specific adaptations.

However, emissions from training are only part of the picture. The operational phase—often referred to as inference—can be even more energy-intensive at scale (Jegham et al., 2025). Every time an LLM generates text, it requires significant real-time computation, multiplied across millions or billions of user queries. A single ChatGPT query is estimated to be up to 1,000 times more energy-intensive than a Google search, according to some analyses,[4] though all such figures remain rough estimates owing to the lack of transparency in the industry. This ongoing energy use contributes to what some researchers call the hidden cost of intelligence as a service: a shift from one-time training to perpetual environmental overhead (Varoquaux et al., 2025).

Importantly, these costs are geographically and politically uneven. Data centers powering the AI revolution are frequently situated in

---

[4] https://tinyml.substack.com/p/the-cost-of-inference-running-the.

regions with cheap electricity and large volumes of freshwater for cooling. This includes rural or semirural areas in Scandinavia, the American Midwest, or parts of South and Southeast Asia, where companies benefit from lower regulatory scrutiny and reduced energy tariffs. In many cases, the electricity is still partially derived from fossil fuels, and the water diverted for data center cooling strains local ecosystems and municipal supplies—sometimes during drought conditions. Local communities often have limited say in these developments, which are shaped by opaque contracts and economic incentives rather than democratic consultation (Li et al., 2023; Siddik, et al., 2021).

In parallel, the race to scale up AI models is driving demand for specialized computing hardware: graphics processing units, tensor processing units, and large-scale server infrastructures. These devices depend on extractive supply chains involving critical and rare earth minerals such as lithium, cobalt, and neodymium. Mining and refining these materials are environmentally destructive processes associated with deforestation, toxic waste, and water pollution. The social costs are also stark: Many mining operations are located in countries across the Global South, where labor protections are weak, child labor is not uncommon, and local communities are displaced or exposed to hazardous conditions. In this sense, the 'intelligence' produced by LLMs is not immaterial but built upon a foundation of ecological degradation and global inequality (Muldoon et al., 2024).

Critically, these environmental harms are largely externalized from the sleek interfaces and frictionless experiences offered by commercial AI platforms. As users, we are rarely aware of the infrastructural footprint behind a seemingly simple query or generated paragraph. This disconnection raises fundamental questions about sustainability, responsibility, and the ethics of scale. While some industry actors have committed to carbon offsets or energy-efficient training methods, such measures are often limited in scope and difficult to verify. Moreover, offsetting does little to address the broader material and geopolitical consequences of rapid AI expansion.

In short, the environmental challenge of LLMs is not just about emissions—it is about energy justice, material accountability, and the uneven distribution of risk and harm across populations and geographies. As the field continues to evolve, a serious reckoning with these ecological dimensions will be essential—not only for the legitimacy of AI research but for its alignment with planetary boundaries and global social equity.

4.2 The hidden labor behind artificial intelligence

Equally important, though often overshadowed by discussions of automation and innovation, are the human labor conditions that underpin the development of LLMs. Despite the popular image of AI as autonomous and self-improving, its foundations remain deeply entangled with vast amounts of invisible human work. Beyond the human workforce required for hardware development as mentioned in the previous section, at nearly every stage of the AI development pipeline—from dataset curation to safety alignment— human labor plays a critical, if largely unacknowledged, role.

A significant portion of this labor is carried out by annotators, content moderators, and data cleaners—tasks sometimes described as 'ghost work' (Gray and Suri, 2019). These workers, often based in low- or middle-income countries, are tasked with labeling datasets, identifying and removing harmful content, generating training prompts, or evaluating outputs for relevance and safety. The conditions of this work are frequently precarious: low wages, piecework pay structures, minimal labor protections, and constant exposure to disturbing material. Workers are also subjected to tight surveillance and algorithmic productivity tracking, with little opportunity for collective bargaining or visibility within the industry.

A stark illustration of this can be found in the widely reported case involving Sama, an outsourcing company contracted by OpenAI via a third-party vendor. In 2022, it was revealed that workers in Kenya were being paid as little as $1.32 to $2 per hour to filter toxic content—including graphic depictions of violence, abuse, and hate speech—to help train and 'align' models like Chat-GPT. Many of these workers reported psychological distress, insufficient mental health support, and a lack of transparency about the nature of their tasks. While OpenAI later ended the contract, the case shed light on the broader outsourcing practices that support the AI industry—practices that persist across multiple companies and continents.

From a philosophical perspective, this invisibility reflects what Fraser (2009) describes as misrecognition: a failure to grant proper social standing, respect, or visibility to certain forms of labor and life. In the context of LLMs, misrecognition operates on both social and ecological levels. The human and environmental costs of AI development are systematically externalized, while the benefits—monetary, symbolic, and infrastructural—are centralized. The value extracted from the labor of invisible workers accrues largely to multinational corporations in the Global North, reinforcing long-standing global hierarchies.

This dynamic has led scholars such as Birhane, Mohamed, and others to describe the phenomenon as computational or data colonialism (Birhane, 2020; Mohamed et al., 2020). Under this framework, AI development reproduces a familiar pattern of extraction and enclosure: Data, labor, and natural resources are sourced from the peripheries, processed through centrally maintained systems, and redeployed in ways that rarely benefit the communities from which they originated. This extractive paradigm not only raises ethical concerns but calls into question the sustainability and justice of the AI industry itself.

To speak of 'intelligent' systems without accounting for the global networks of undervalued human labor that sustain them is to perpetuate a myth. Any serious ethics of AI must reckon with these social, psychological, and economic conditions—lest we reproduce the very inequalities that many claim AI has the potential to address. Responding to these externalities requires more than technical optimization. What is needed is a shift in ethical attention—from model performance to system-wide impact, from abstract fairness to *relational justice*. This includes rethinking procurement practices, ensuring fair labor standards in the data supply chain, and developing participatory governance frameworks that include workers, affected communities, and environmental advocates in decisions about AI deployment.

5 Economic power and platform dependence

The contemporary landscape of LLMs is shaped not only by technical capacity or ethical aspiration but also by powerful economic dynamics. The training, fine-tuning, and deployment of frontier models require immense computational infrastructure, specialized expertise, and access to proprietary datasets—resources that are overwhelmingly concentrated in a small number of corporate actors. This concentration has profound implications for the distribution of innovation, agency, and economic value in the age of AI.

5.1 The economics of scale and the privileging of size

The dominant narrative surrounding LLMs often celebrates the continual increase in scale—models with ever more parameters, trained on ever larger datasets, and optimized across a growing array of downstream tasks. This enthusiasm for scaling is frequently portrayed as the natural trajectory of progress, with larger models presumed to be more powerful, more general, and ultimately more valuable. However, this assumption is increasingly contested. Recent research suggests that smaller, task-specific models—particularly when fine-tuned on well-curated data—can outperform larger general-purpose models in specialized domains (Bamman et al., 2024; Varoquaux et al., 2025). These smaller models are often more interpretable, more efficient, and better aligned with concrete human needs. Crucially, they require far less computational energy and hardware to train and deploy, making them more sustainable both ecologically and institutionally.

The current bias toward scale is not driven solely by scientific evidence but by a confluence of economic, infrastructural, and reputational incentives. Training models like GPT-4 or Gemini Ultra entail costs running into the tens or hundreds of millions of dollars, along with access to proprietary datasets, specialized hardware, and highly skilled engineering teams. These costs can be borne only by a small number of firms—primarily based in the United States (most notably OpenAI, backed by Microsoft; Google DeepMind; Meta; and Anthropic, backed by Amazon) and China (where major players include Baidu, Alibaba, Tencent, Huawei, and, more recently, DeepSeek), which currently dominate the field. However, strong alternatives have begun to emerge elsewhere, including Mistral (France), Aleph Alpha (Germany), Cohere (Canada), and the AI2-backed OLMo project (United States, but fully open and research-led). These companies control many of the critical layers of the AI development pipeline—ranging from model training to distribution platforms. As a result, the perception of progress becomes tightly coupled to corporate capacity, rather than to collective benefit or pluralistic inquiry.

This concentration has far-reaching implications. As Joseph Stiglitz (2012) has argued, economic inequalities are often not accidental but the result of deliberate policy choices and institutional arrangements that benefit dominant actors. In the context of LLMs, a similar dynamic is visible: The technological frontier is increasingly shaped by the interests and capabilities of powerful firms, rather than by open scientific deliberation or broad public accountability. The assumption that 'bigger is better' fosters a winner-takes-all dynamic in which smaller entities—universities, public labs, nonprofits, or local governments—find it difficult to secure funding or legitimacy unless they conform to the dominant logic of scale. In this environment, scale becomes both a technical and epistemic filter: Models that are not large enough are often dismissed as inherently less capable, regardless of their actual performance, contextual relevance, or ethical tractability.

Moreover, the opacity that accompanies many large-scale proprietary models exacerbates this asymmetry. Without access to the training data, modeling assumptions, or evaluation procedures, external researchers are unable to reproduce, audit, or contest the claims made about model capabilities. This is not merely a transparency concern but a deeper epistemic problem: As knowledge production becomes embedded in closed, corporate infrastructures, the authority to define what counts as credible reasoning or valid information is increasingly ceded to private actors. In such a landscape, alternative approaches—such as leaner, transparent, domain-specific models—are not just viable but necessary. They offer a path toward more democratic, accountable, and sustainable forms of machine intelligence.

5.2 Platform lock-in and technofeudalism

Despite concerns over concentration and opacity, it is crucial to recognize that the widespread adoption of corporate LLM platforms is not simply imposed but actively chosen—often for good reason. These systems are reliable, user-friendly, integrated across devices and workflows, and backed by extensive support infrastructures. For many educators, developers, and businesses, they offer a low-friction path to innovation and experimentation. Moreover, they benefit from network effects: The more people use a given platform, the more refined its outputs become, and the more compelling it is to remain within its ecosystem.

This creates a dynamic of platform lock-in, in which users become increasingly dependent on a particular model or interface—not because alternatives are unavailable but because the cost of switching (in terms of compatibility, usability, or performance) becomes prohibitive. As with earlier shifts in software and cloud infrastructure, the result is a subtle but powerful form of technological enclosure: Cognitive labor, linguistic creativity, and even personal memory are routed through proprietary channels, subject to the terms and conditions of a few corporations.

From a philosophical perspective, this situation raises questions about autonomy and structural dependence. As Foucault (1978) emphasizes, power is not only coercive but productive: It shapes what is possible, desirable, and thinkable within a given system. In the case of LLMs, the productivity of corporate AI platforms— their capacity to generate text, suggest actions, or summarize complex information—comes at the cost of embedding users within architectures they do not control. This is not inherently exploitative; it also limits the horizon of choice and agency in subtle ways.

More recently, Yanis Varoufakis (2024) has provocatively argued that the economic logic of the digital age no longer fits traditional models of capitalism (following a theory originally proposed by Durand, 2020, for the digital economy in general). In classical capitalist frameworks—whether Adam Smith's competitive markets or Marx's account of industrial production—economic power is organized around the ownership of capital and the exploitation of labor in the production of commodities. Value is generated through exchange in relatively open markets, where firms compete for profit and consumers exercise choice, however constrained. Even under monopoly capitalism, where competition is reduced, the basic logic of market mediation remains central: Goods and services are bought and sold, labor is waged, and surplus value is extracted in the factory or the firm.

Technofeudalism, by contrast, designates a new political-economic formation in which value is no longer primarily generated through market exchange but through the enclosure and control of digital territories. In Varoufakis' terms, we are witnessing the rise of 'cloud capitalists'—a handful of companies that own and operate the computational infrastructure, data pipelines, and algorithmic systems that mediate digital life. These actors primarily compete no longer by producing goods but by governing access: controlling platforms, setting terms of service, and extracting rent from users who operate within walled gardens. Users are no longer consumers in a classical market sense but platform-dependent tenants—locked into proprietary ecosystems, unable to exit without losing data, identity, or functionality.

This model bears a structural resemblance to feudalism: not in the literal sense of lords and vassals, but in the underlying logic of jurisdictional control over bounded domains. Like feudal lords who granted access to land in exchange for tribute or loyalty, cloud firms grant access to digital spaces—such as social networks, content platforms, and, now, AI interfaces—while retaining ultimate sovereignty over the rules, infrastructure, and extracted value. Importantly, the 'rent' extracted is not only monetary but also behavioral, cognitive, and epistemic: User interactions, queries, corrections, and preferences feed back into the model, improving the system's performance while consolidating the firm's advantage.

In the context of LLMs, this dynamic is especially pronounced. When users interact with proprietary LLMs, they contribute to the refinement of the system—often without explicit compensation or recognition. The value produced through these interactions accrues to the platform provider, while users remain tethered to interfaces they do not control, and whose inner workings they cannot inspect. This is a shift not merely in business models but in the political economy of cognition: Linguistic labor, creative expression, and even memory become resources to be harvested within enclosed digital territories.

Technofeudalism thus reframes debates about innovation and competition. The problem is not simply that a few firms are 'too big' or that regulation lags behind. It is that the very structure of participation in digital life has been reorganized around asymmetrical control and nonmarket dependency. Unlike traditional markets, where exit is at least theoretically possible, cloud-based AI ecosystems rely on high switching costs, proprietary formats, and network effects that make departure difficult or self-defeating. The result is a form of cognitive capture: a situation in which individuals and institutions must operate within the epistemic parameters defined by a few platform sovereigns, whose interests may not align with public goods or democratic norms (Couldry and Mejias, 2019).

Whether or not one accepts Varoufakis's full diagnosis, it highlights an important shift in the locus of economic control. In the world of LLMs, users generate value for platforms by interacting with them: providing prompts, corrections, and behavioral data that continuously refine the model. This creates a feedback loop of unpaid digital labor—similar to what Terranova (2000) earlier described as 'free labor' in online platforms. The difference is that the outputs of this labor are now folded into systems of synthetic cognition, whose benefits are unevenly distributed and largely unaccountable to the communities from which their value is drawn (e.g., users, annotators, data contributors).

The risk here is not only economic concentration but epistemic capture. As corporate platforms become the default interface for writing, searching, reasoning, and remembering, they shape not only what users can do but what they can imagine doing. The promise of LLMs as a democratizing force is thus in tension with their role as gatekeepers: They offer access but only through systems that reinforce their own centrality.

5.3 Toward a pluralist political economy of language models

Confronting these dynamics does not require rejecting LLMs or demonizing the companies that have played a leading role in their development. The issue is not the existence of corporate actors per se but the concentration of epistemic and infrastructural power in a small set of institutions whose priorities are shaped by profit motives, proprietary logic, and private governance. A critical response, then, must go beyond calls for transparency or 'ethical AI' and instead target the underlying structures that shape who controls these technologies, how they are developed, and in whose interests they operate.

What is needed is a reorientation of both governance frameworks and incentive structures: mechanisms that actively support the development of alternative models and infrastructures. This includes public investment in computational resources, research programs focused on open and responsible AI, and legal safeguards that limit excessive enclosure or vertical integration across the AI value chain. Without such systemic interventions, the trajectory of

LLM development will remain aligned with the interests of a few firms, regardless of broader social or epistemic consequences.

A promising avenue lies in the creation and long-term support of public, civic, or cooperative LLMs—models built and maintained by universities, independent research labs, governmental institutions, or consortia of nonprofit organizations. These initiatives can embed democratic values into the design and deployment of language technologies, not as external constraints but as integral principles. Examples we have already evoked, such as Bloom (developed by BigScience), OLMo (from the Allen Institute for AI), and Poro (by TurkuNLP), demonstrate that high-performing, fully open models can be built outside the confines of corporate silos. These projects emphasize not only open access but also transparent documentation, reproducibility, and community governance, helping to reconfigure the meaning of responsibility in AI development.

However, these efforts remain fragile and under-resourced in comparison to their corporate counterparts. A model like Bloom required massive volunteer coordination and relied heavily on institutional goodwill, while still lacking long-term maintenance funding. OLMo takes an important step by combining open weights with open training data and code, but its sustainability depends on continued philanthropic support. If such models are to offer a serious alternative, they will require more than symbolic endorsement—they need institutional embedding, stable funding, and policy protections that shield them from predatory acquisition or obsolescence.

The normative case for this shift is underscored by the work of political philosophers such as Anderson (2017), who argues for the extension of economic democracy—the idea that institutions with profound effects on people's lives should be governed by, or at least accountable to, those people. When applied to AI, this principle implies that governance should not be left to the invisible hand of market demand or the closed deliberations of technical experts. Instead, it requires mechanisms for collective deliberation, worker and user representation, and accountability to public values—whether through state regulation, cooperative models, or transnational frameworks of algorithmic oversight.

This shift also challenges dominant narratives about 'innovation' as a product of entrepreneurial freedom and rapid scaling. A truly democratic approach to LLMs would recognize that innovation can—and should—occur in slower, more deliberative contexts that prioritize social utility, linguistic inclusivity, and long-term epistemic integrity over market share. Rather than asking which firms will 'win' the AI race, we should ask what kinds of knowledge infrastructures we want to live with, and how they can serve pluralistic societies.

6 Conclusion: living ethically with large language models

If LLMs are no longer confined to research labs but embedded in the textures of everyday life—structuring knowledge, shaping communication, influencing decisions—then the question becomes not simply how to govern or constrain them but how to *live with them*. This final section turns from critique to possibility, asking how we might cultivate ethical relationships with LLMs—relationships that preserve human agency, foster diversity, and foreground care.

Philosophers of technology have long argued that ethical life cannot be reduced to compliance with external norms; it involves *practical wisdom*—the capacity to navigate complex, situated dilemmas with sensitivity and discernment. In this light, ethical interaction with LLMs is not a matter of obeying rules but one of developing reflexive habits: pausing before accepting a model's suggestion, questioning the assumptions behind its output, considering who or what might be rendered invisible in its formulation. As Vallor (2016) argues, we need not only technomoral rules but *technomoral virtues*—qualities like humility, attentiveness, and responsibility, cultivated over time through use.

This is especially important given the tendency of LLMs to normalize certain linguistic patterns, cultural frames, or moral intuitions through repetition. Over time, users may come to accept what is statistically probable as what is ethically or socially desirable. Such convergence is subtle but powerful; it shapes what seems 'natural' to say, ask, or believe. Living ethically with LLMs means resisting this drift—remaining open to plurality, dissensus, and the unanticipated. It means noticing when a model flattens nuance, erases ambiguity, or offers resolution where real ethical deliberation would require discomfort or delay.

From a care ethics perspective, this entails rethinking our relation to technology not as domination or control but as stewardship—a reciprocal, attentive engagement that recognizes vulnerability and interdependence (Held, 2006). LLMs are not moral agents but they are objects of moral concern: They are made by human hands, reflect human worlds, and mediate human relations. Caring for them means caring about *how* they are built, *whom* they serve, and *what* they make possible or foreclose. This perspective encourages us to value slowness over speed, contextuality over generalization, and cocreation over automation.

Practically, living ethically with LLMs might involve developing new institutional and pedagogical frameworks. Educational curricula should equip students not only to use LLMs effectively but also to *read* them critically—to recognize their biases, limitations, and cultural positions. Designers and developers might adopt practices of 'ethical prototyping' (Verbeek, 2011), embedding value-sensitive reflection into every stage of system development. And publics should be invited into deliberation—not only about AI regulation but also about the imaginaries that shape its trajectory: What do we want language technologies to do? What kinds of relations do we want them to foster?

References

Anderson, E. (2017). *Private government: How employers rule our lives (and why we don't talk about it)*. Princeton University Press.

Bamman, D., Chang, K. K., Lucy, L., and Zhou, N. (2024). On classification with large language models in cultural analytics. In *CHR 2024:*

Computational Humanities Research Conference, Aarhus, Denmark. https://arxiv.org/abs/2410.12029

Birhane, A. (2020). Algorithmic colonization of Africa. *SCRIPTed*, *17*(2), 389. https://doi.org/10.2966/scrip.170220.389

Bowker, G. C., and Star, S. L. (1999). *Sorting things out: Classification and its consequences.* MIT Press.

Chen, Z., Ma, J., Zhang, X., Hao, N., Yan, A., Nourbakhsh, A., Yang, X., McAuley, J., Petzold L. R., and Wang, W. Y. (2024). A survey on large language models for critical societal domains: Finance, healthcare, and law. *Transactions on Machine Learning Research.* https://arxiv.org/abs/2405.01769

Choudhury, A. and Chaudhry, Z. (2024). Large language models and user trust: Consequences of self-referential learning loop and the deskilling of health care professionals. *Journal of Medical Internet Research, 26.* https://www.jmir.org/2024/1/e56764/

Citron, D. K., and Pasquale, F. (2014). The scored society: Due process for automated predictions. *Washington Law Review, 89*(1), 1–33.

Couldry, N., and Mejias, U. A. (2019). *The costs of connection: How data is colonizing human life and appropriating it for capitalism.* Stanford University Press. https://doi.org/10.1515/9781503609754

Dreyfus, H. L. (1992). *What computers still can't do: A critique of artificial reason.* MIT Press.

Durand, C. (2020). *How Silicon Valley unleashed techno-feudalism: The making of the digital economy.* Verso.

Foucault, M. (1978). *The history of sexuality, Volume 1: An introduction* (R. Hurley, Trans.). Pantheon Books.

Fraser, N. (2009). *Scales of justice: Reimagining political space in a globalizing world.* Columbia University Press.

Fricker, M. (2007). *Epistemic injustice: Power and the ethics of knowing.* Oxford University Press.

Gray, M. L., and Suri, S. (2019). *Ghost work: How to stop Silicon Valley from building a new global underclass.* Houghton Mifflin Harcourt.

Held, V. (2006). *The ethics of care: Personal, political, and global.* New York: Oxford University Press.

Ihde, D. (1990). *Technology and the lifeworld: From garden to earth.* Indiana University Press.

Jegham, N., Abdelatti, M., Elmoubarki, L., and Hendawi, A. (2025). How hungry is AI? Benchmarking energy, water, and carbon footprint of LLM inference. arXiv preprint. https://arxiv.org/abs/2505.09598

Kant, I. (1996 [1784]). An answer to the question: What is Enlightenment? In J. Schmidt (Ed.), *What is Enlightenment? Eighteenth-century answers and twentieth-century questions* (pp. 58–64). University of California Press.

Latour, B. (1992). Where are the missing masses? The sociology of a few mundane artifacts. In W. E. Bijker and J. Law (Eds.), *Shaping technology/building society: Studies in sociotechnical change* (pp. 225–58). MIT Press.

Li, P., Yang, J., Islam, M. A., and Ren, S. (2023). Making AI less thirsty: Uncovering and addressing the secret water footprint of AI models. arXiv preprint. https://arxiv.org/abs/2304.03271

Liu, Z. (2024). Cultural bias in large language models: A comprehensive analysis and mitigation strategies. *Journal of Transcultural Communication*, 3(2). https://doi.org/10.1515/jtc-2023-0019

Mohamed, S., Png, M.-T., and Isaac, W. (2020). Decolonial AI: Decolonial theory as sociotechnical foresight in artificial intelligence. *Philosophy & Technology*, 33, 659–684. https://doi.org/10.1007/s13347-020-00405-8

Muldoon, J., Graham M., and Cant, C. (2024). *Feeding the machine. The hidden human labour powering AI*. Canongate Books.

Patterson, D., Gonzalez, J., Le, Q., Liang, C., Munguia, L. M., Rothchild, D., So, D., Texier, M., and Dean, J. (2021). Carbon Emissions and Large Neural Network Training. arXiv preprint. https://arxiv.org/abs/2104.10350

Siddik, A. B., Shehabi, A., and Marston, L. (2021). The environmental footprint of data centers in the United States. *Environmental Research Letters*, 16(6), 064017.

Stiglitz, J. E. (2012). *The price of inequality: How today's divided society endangers our future*. W. W. Norton & Company.

Strubell, E., Ganesh, A., and McCallum, A. (2019). Energy and policy considerations for deep learning in NLP. In *Proceedings of the 57th Annual Meeting of the Association for Computational Linguistics (ACL), Florence, Italy* (pp. 3645–3650). Association for Computational Linguistics. https://aclanthology.org/P19-1355/

Suchman, L. A. (2007). *Human–machine reconfigurations: Plans and situated actions* (2nd ed.). Cambridge University Press.

Taylor, L. (2017). What is data justice? The case for connecting digital rights and freedoms globally. *Big Data & Society*, 4(2). https://journals.sagepub.com/doi/full/10.1177/2053951717736335

Terranova, T. (2000). Free labor: Producing culture for the digital economy. *Social Text*, 18(2), 33–58.

Vallor, S. (2016). *Technology and the virtues: A philosophical guide to a future worth wanting.* Oxford University Press.

Varoquaux, G., Luccioni, A. S., and Whittaker, M. (2025). Hype, sustainability, and the price of the bigger-is-better paradigm in AI. In *ACM Conference on Fairness, Accountability, and Transparency (FAccT).* Preprint: https://arxiv.org/abs/2409.14160

Varoufakis, Y. (2024). *Technofeudalism: What killed capitalism.* Melville House.

Verbeek, P.-P. (2011). *Moralizing technology: Understanding and designing the morality of things.* University of Chicago Press.

CONCLUSION

Thinking with machines

As we reach the end of this inquiry into large language models (LLMs), a broader picture begins to emerge—one that returns us to the notion, raised in the introduction, of these systems as "philosophical provocations' (McGinn, 2017). They do not offer final answers to long-standing questions about language, mind, or meaning. Instead, they act as conceptual stress tests, revealing the fragility of inherited distinctions and unsettling the boundaries between simulation and understanding, use and reference, fluency and knowledge. What follows is not a summary but a reflection on the deeper implications of LLMs for how we think about thinking itself.

The three final sections consider, in turn, the theoretical disruptions that LLMs provoke, the cognitive comparisons they invite, and the hard epistemic limits they expose. Rather than attempting to resolve the tensions these systems bring to light, this conclusion dwells on them—treating them not as problems to be solved but as openings for reimagining what intelligence, both human and artificial, might be.

How to cite this book chapter:
Poibeau, T. 2025. *Understanding Conversational AI: Philosophy, Ethics and Social Impact of Large Language Models.* Pp. 229–236. London: Ubiquity Press. DOI: https://doi.org/10.5334/bde.k. License: CC BY-NC 4.0

1 How LLMs disrupt established theories

Among the many provocations posed by LLMs, one of the most consequential is their capacity to unsettle established theories without offering clear alternatives. Rather than proposing new doctrines of language, mind, or knowledge, LLMs function in ways that challenge the coherence or sufficiency of long-held assumptions. They compel us to ask whether concepts like understanding, meaning, or grammar—as traditionally defined—are adequate to account for what these systems do. Their significance lies not in resolving theoretical debates but in revealing where existing frameworks may have rested on too narrow a view of cognition, communication, or intelligence. In this sense, LLMs act less as answers than as 'epistemological irritants,' forcing a reconsideration of the terms in which foundational questions have been posed.

This is especially clear in the case of understanding. Philosophically, the notion of understanding has often been tied to grounding: the capacity to refer to objects, to evaluate statements for their truth, or to situate utterances within a broader conceptual or perceptual framework (Frege 1993 [1892]; Putnam 1975; Searle 1980). LLMs do none of these things. They do not refer, they do not believe, they do not perceive—at least not in the traditional sense, since their only source of knowledge is text. Yet they produce context-sensitive, coherent responses to a wide array of prompts. Do they 'understand'? The question turns out to be less about the models than about our own conceptual habits. If understanding requires explicit, external grounding, then no. But if understanding is redefined functionally—as the capacity to use language appropriately across contexts—then the answer may be yes, or at least: not entirely no. In a way, LLMs bring to life the poststructuralist conception of discourse—as articulated by Barthes and Foucault—as fragmented, authorless, testing its implications in a computational context that earlier theorists could only imagine.

A similar destabilization occurs with respect to meaning. Philosophers and linguists have offered an array of intricate theories about what meaning is: referentialist, inferentialist, conceptual-role based, formalist, use-based. Most of these theories take for granted that meaning is something to be explained. LLMs reverse this perspective. They do not seek to explain meaning—they model it through use. Meaning, for them, is what emerges when words are placed in context. There is no ontology, no dictionary, no mapping to an external world (at least, with 'pure' language models—we saw in Chapter 1 that retrieval-augmented generation, human feedback, and external knowledge bases are ways to go beyond this limitation of the original models). Instead, meaning becomes a function of linguistic cooccurrence, an effect of distributional regularity. This brings computational force to a view that has long remained on the margins of mainstream semantic theory: That meaning is not a fixed entity but a pragmatic function, shaped by usage, genre, and social practice.

Nowhere is this reconfiguration more evident than in the status of grammar. In generative linguistics, grammar has typically been understood as a set of formal rules or constraints, often universal in scope, encoded (implicitly or explicitly) in the minds of speakers. But LLMs do not encode rules. They are not provided with grammars. Instead, they acquire the ability to produce grammatically correct text simply by learning statistical associations. Syntax, in this context, is not a precondition for generation—it is an emergent property of successful prediction. This empirical inversion challenges deeply held assumptions about the modularity and innateness of syntactic knowledge, and it aligns more closely with usage-based and connectionist models of language acquisition, in which structure emerges from exposure rather than being imposed a priori. The analogy with child language learning is striking: Children acquire the capacity to communicate long before they are aware of grammatical rules. Syntax is not taught—it is inferred. The same is true for LLMs. In both cases, grammar is not a cause but an effect.

These insights have consequences beyond linguistics. In philosophy of mind, notions such as theory of mind, consciousness, and intentionality are increasingly difficult to define in the face of models that exhibit behavior previously thought to require such capacities. If a system can simulate the behavior of an agent with beliefs, does it 'have' a theory of mind—or does the concept itself need to be revised to accommodate simulated interactivity without mental representation? In psychology, too, LLMs challenge the distinction between symbolic and subsymbolic processing, between rule-following and pattern recognition, between 'true' cognition and its mimetic shadow. The models provide a concrete instantiation of distributed, nonexplicit learning—suggesting that some aspects of cognition may be more statistical, more ecological, and less modular than traditionally assumed.

What emerges from all of this is not the displacement of philosophy or linguistics but a renewed opportunity for reflection. LLMs do not replace existing theories but they reveal some of the assumptions those theories have taken for granted. They bring to light possibilities that our current conceptual frameworks are still learning to accommodate. In doing so, they encourage a more pragmatic, experimental, and plural approach to questions of meaning, language, and mind. If we are open to learning with them—not just about them—we may end up reconsidering not only how we build machines but also how we make sense of ourselves.

2 Beyond correlation in artificial and human intelligence

A recurring question in discussions about LLMs is whether humans, like these systems, are fundamentally pattern matchers. On the surface, LLMs appear to mirror key aspects of human cognition: They learn from examples, generalize across contexts, and produce meaningful outputs without relying on explicit rule-based representations. However, closer inspection reveals that this resemblance conceals important differences—both in how these systems operate and in what their limitations expose about the nature of thinking itself.

LLMs such as ChatGPT are trained on vast corpora of text, learning to generate plausible continuations based on statistical regularities. This allows them to simulate a wide range of discourses, including those that involve abstract reasoning. In many instances, they perform impressively well: They can imitate mathematical proofs, summarize philosophical arguments, and produce syntactically sophisticated prose. But the underlying mechanism remains the same—next-word prediction based on probabilistic associations. In this sense, their capabilities are grounded in correlation rather than in comprehension.

This limitation becomes especially apparent when we turn to tasks such as arithmetic. For a surprisingly long time, earlier versions of ChatGPT struggled with basic multiplication involving large numbers. The system would produce answers that were approximate—close in magnitude, with some correct digits—but still incorrect. The result was not the outcome of faulty logic but of the absence of logic altogether. No actual multiplication was being performed; rather, the model was generating plausible-seeming numerical sequences based on patterns seen during training.

This is not a trivial patch. It reveals something fundamental about the limits of inference from examples. Arithmetic is extremely difficult—if not impossible—to master solely by observing many instances. Human children do not learn multiplication just by watching others multiply; they are taught explicit procedures, internalize abstract rules, and come to understand the structure of number systems in ways that support generalization far beyond their direct experience; '4' is not similar to '5' in the way that 'strong' is similar to 'powerful.' Some domains require more than statistical association—they depend on formal, symbolic operations. This suggests that not all cognition is reducible to pattern recognition, and that there are more economical and structured paths to understanding than mere exposure to billions of examples.

This distinction has broader implications. It reminds us that intelligence—whether human or artificial—is not monolithic. There is no single way to generalize, no universal mode of reasoning that captures all forms of understanding. The capabilities of LLMs demonstrate the power of large-scale pattern recognition, but also its limits. They can convincingly simulate mathematical reasoning but, without mechanisms for verification or formal manipulation, they are prone to errors that a basic calculator would never make (ChatGPT, which long struggled with basic calculations, finally integrated a calculator in 2024, effectively resolving the problem). To extend their abilities, it is necessary to augment pattern-based inference with other forms of reasoning: symbolic computation, external tools, domain-specific knowledge, and perhaps, eventually, embodied interaction with the world (something LLMs already accommodate through their ability to use tools and operate as agents; see Section 8 of the technical annex).

These limitations are not failures; they are informative boundaries. They highlight that what we call 'intelligence' consists of multiple modes of processing—some inferential, some procedural, some experiential—and that artificial systems, like humans, may need to integrate diverse strategies to navigate complex tasks. The dream of artificial general intelligence often presumes a unified capacity to generalize across domains. But the example of LLMs suggests otherwise: that intelligence may be irreducibly plural, and that different problems call for different kinds of minds.

3 Uncertain futures: societal challenges at the scale of LLMs

If LLMs challenge long-standing theories of language, cognition, and meaning, they also confront society with a diverse and deeply consequential set of practical challenges. These are not only technical or philosophical but societal in the broadest sense—affecting how we learn, how we allocate resources, and how we make collective decisions. The future of LLMs is therefore not only a matter of engineering but also one of education, environment, and ethics.

One pressing challenge concerns the changing nature of learning. As LLMs become integrated into classrooms, search engines, and writing tools, they alter how knowledge is accessed and how intellectual labor is distributed. Will these systems serve to enhance understanding or will they encourage dependence on surface-level outputs? There is a risk that automated fluency may come to substitute for deep comprehension, especially when generative tools produce plausible but unverifiable responses. At the same time, these tools could support more personalized, adaptive forms of learning—if used judiciously. Much depends on how we design educational environments that incorporate such technologies without eroding the value of critical thinking, creativity, and conceptual rigor.

At the other end of the spectrum lies a challenge of planetary scope: the environmental cost of large-scale computation. Training and deploying LLMs requires enormous energy and material resources. Their carbon footprint is substantial, and their appetite for data and compute power shows little sign of slowing down (see Chapter 9). As climate change accelerates, this model of scale-as-progress becomes increasingly difficult to justify. The development of more efficient architectures is a technical priority but efficiency alone will not resolve the deeper contradiction between the pursuit of ever more powerful models and the urgent need for ecological sustainability. The compatibility of LLMs with long-term planetary preservation remains an open question.

Finally, LLMs raise profound issues of governance and epistemic equity. These models are trained on vast, heterogeneous corpora that reflect existing cultural, linguistic, and social hierarchies. Yet their development and control remain concentrated in a handful of institutions and geopolitical regions. This raises concerns not only about economic power and data ownership but also about whose knowledge is encoded, whose voices are marginalized, and who sets the norms for what counts as a valid or valuable output. As LLMs become embedded in legal, medical, and educational systems, ensuring transparency, accountability, and inclusivity will become increasingly vital.

Taken together, these challenges—ranging from education to environmental impact to epistemic justice—underscore the societal scale of the questions posed by LLMs. They are not isolated side effects but central dimensions of what it means to live with and through intelligent systems. While much about the future of

LLMs remains uncertain, one thing is clear: The choices we make now will shape not only how these models evolve but how we learn to live with them.

References

Frege G. [1892 (1993)]. On sense and reference. In A. W. Moor (Ed.), *Meaning and reference* (pp. 23–42). Oxford University Press.

McGinn, C. (2017). *Philosophical Provocations*. The MIT Press.

Putnam, H. (1975). Language, mind, and knowledge. *Minnesota Studies in the Philosophy of Science, 7*, 131–193. Retrieved from the University Digital Conservancy, https://hdl.handle.net/11299/185225

Searle, J. (1980). Minds, brains and programs. *Behavioral and Brain Sciences, 3*(3), 417–457. https://doi.org/10.1017/S0140525X00005756

The architecture and training of large language models

1 Introduction

This technical annex provides a comprehensive, nonnormative account of how large language models (LLMs) function, from the representation of textual data to the architectures, training processes, and alignment techniques that enable their deployment as interactive systems.

LLMs such as GPT, Llama, Gemini, Mistral, and Claude, are based on a shared foundation: the transformer architecture introduced by Vaswani et al. (2017). These models are trained on massive corpora of text using a self-supervised learning objective, typically causal language modeling. Following pretraining, foundation models can be further aligned with human instructions through techniques such as supervised fine-tuning (SFT) and reinforcement learning from human feedback (RLHF). This annex focuses on the technical components that underpin such models.

Its primary goal is to serve as a reference for readers of the main volume, which addresses philosophical, ethical, and sociopolitical

How to cite this book chapter:
Poibeau, T. 2025. *Understanding Conversational AI: Philosophy, Ethics and Social Impact of Large Language Models.* Pp. 237–252. London: Ubiquity Press. DOI: https://doi.org/10.5334/bde.l. License: CC BY-NC 4.0

questions raised by LLMs. Here, we abstract away from ethical and social concerns and focus exclusively on architectural and computational mechanisms.

The annex is structured as follows. Section 2 introduces how language is represented numerically through tokenization and embedding. Section 3 explains the architecture of transformers, including self-attention and model scaling. Section 4 describes training objectives, including the distinction between training and inference. Section 5 discusses empirical scaling laws. Section 6 details the transition from foundation models to instruction-following chat models, including RLHF. Section 7 outlines control mechanisms and safety filters. Sections 8 and 9 address emerging augmentations and current limitations. A bibliography is provided at the end.

2 Text representation: from language to numbers

2.1 Tokenization

Before any text can be processed by a language model, it must be transformed from a sequence of characters into discrete units known as tokens. Tokenization is the process of segmenting text into these tokens, which serve as the model's input.

In early natural language processing systems, tokenization was often performed at the word level. However, modern LLMs typically use subword tokenization strategies, such as byte pair encoding (BPE), WordPiece, or SentencePiece. These methods allow rare or out-of-vocabulary words to be decomposed into smaller, more frequent components. For example, the sentence

The revolution will not be televised.

might be tokenized as:

['The', 're', 'vol', 'ution', 'will', 'not', 'be', 'tele', 'vised', '.']

BPE merges frequent character pairs, WordPiece selects subwords to maximize likelihood in a language model, and SentencePiece treats input as a raw byte or character stream and uses unsupervised algorithms (like BPE) to learn subword units, making it possible to process all kinds of languages, including those written in nonalphabetic scripts.

Although this segmentation may seem awkward—since it does not follow standard linguistic rules—it enables the use of a fixed and limited vocabulary of tokens, despite the quasi-infinite number of possible word forms found on the web. This balances vocabulary efficiency with flexibility, enabling the model to handle rare, compound, or novel words robustly. However, the impact of tokenization on subsequent model performance is not yet fully understood and remains an active area of research.

2.2 Vectorization

Once tokenized, the input must be converted into numerical form. Each token is mapped to a fixed-size vector through an embedding matrix. These vectors are high-dimensional (e.g., 512, 1024, or 4096 dimensions) and are learned during training.

The embedding process indirectly captures relationships between tokens: Those that appear in similar contexts tend to share similar vector representations. For example, words like 'king' and 'queen' or 'apple' and 'banana' are often close in the embedding space. In contrast, 'banana' and 'king' typically occupy distant regions owing to their unrelated meanings.

By default, vectors do not take into account the order of tokens in a sequence. These static embeddings do not capture ambiguity or polysemy: A word like 'bank' will have only one representation, conflating in a single vector its distinct meanings, such as a financial institution and the side of a river.

3 Core architecture: the transformer

The transformer architecture is the foundation of most contemporary LLMs, including systems such as ChatGPT, Claude, and Mistral. Introduced by Vaswani et al. (2017), the transformer departed from earlier approaches based on recurrence or convolution, instead relying on a mechanism known as self-attention. This innovation enabled much greater efficiency in training and a significantly improved ability to model complex patterns in language.

Unlike previous architectures that processed words one at a time or in fixed sequences, the transformer allows the model to consider all tokens in a sequence simultaneously. This parallelism facilitates the identification of long-range dependencies and nuanced contextual relationships within a text.

3.1 Self-attention mechanism

Self-attention is a central feature of the transformer architecture. It enables the model to evaluate the relevance of each token in a sequence relative to every other token. This allows the model to account for context and capture complex linguistic features such as coreference and syntactic structure.

For instance, in the sentence 'The dog chased the ball because it was fast,' the model must infer whether 'it' refers to the dog or the ball. Through self-attention, the model assigns weights to the other tokens in the sentence, indicating how much influence each one should have when computing a new representation for the token 'it.' These weights are learned functions that reflect the contextual similarity between tokens.

Modern transformers employ multihead attention, which runs several self-attention mechanisms in parallel. Each head can capture a different type of linguistic or semantic relation, thereby enriching the model's ability to represent complex and overlapping language patterns.

3.2 Feedforward layers, residual connections,
and normalization

Each transformer block also includes a feedforward layer—a small neural network that processes the output of the attention mechanism for each token individually. This network identifies more abstract features of the input, such as syntactic roles or semantic categories.

To enable the training of very deep models, transformers make use of residual connections. Rather than replacing the token representation at each step, the model adds the new information to the original input. This design improves gradient flow and preserves useful features during training.

In addition, layer normalization is applied to stabilize the values passed through the network and to facilitate more efficient learning. Normalization ensures that the scale of the input remains consistent across layers, preventing numerical instability and speeding up convergence.

By stacking many such layers—often dozens or even hundreds—transformers construct hierarchical representations of language. These representations allow the model to integrate both local features (e.g., morphology and syntax) and broader patterns (e.g., discourse structure or narrative coherence).

3.3 Model depth, width, and scalability

The capacity of a transformer model is determined by several factors: its depth (the number of stacked layers), its width (the dimensionality of internal representations), and the number of attention heads used per layer. Recent models such as GPT-4, Claude 3 Opus, and Mistral's Mixtral employ deep and wide configurations, with many layers, thousands of hidden dimensions, and numerous parallel attention heads.

Although the exact specifications of these models are not always publicly disclosed, it is known that they incorporate advanced scaling techniques, including 'mixture of experts' architectures. These allow the model to activate only a subset of its parameters for a given input, thereby improving computational efficiency and scalability.

Such design choices enable LLMs to learn long-range dependencies, subtle semantic distinctions, and rich linguistic patterns. However, scaling these models also dramatically increases the demand for training data and computational resources, raising significant questions regarding their environmental and economic sustainability.

4 Training large language models

4.1 Training objective

LLMs are typically trained using a self-supervised learning objective known as causal language modeling (CLM). In this setup, the model is given a sequence of tokens and is trained to predict the next token in the sequence. This is achieved by minimizing the cross-entropy loss between the predicted distribution and the actual next token.

Formally, for a token sequence $t_1, t_2, ..., t_n$, the model maximizes the likelihood:

$$P(t_1, t_2, ..., t_n) = \Pi_i\, P(t_i \mid t_1, ..., t_{i-1})$$

This autoregressive training allows the model to learn probabilistic dependencies across arbitrarily long sequences, subject to the limits of the attention window. Unlike masked language modeling (used in BERT), which predicts randomly masked tokens from their context, CLM only sees past tokens, not future ones.

4.2 Pretraining data and compute

Training an LLM requires large-scale corpora, typically comprising hundreds of billions to trillions of tokens. These datasets are drawn from diverse sources such as Common Crawl, Wikipedia, books, academic papers, and code repositories (or just the Web). Some models use filtered datasets to reduce low-quality content, duplicates, or toxic language.

Training a large-scale LLM also demands significant computational resources. Models like GPT-3 were trained on thousands of GPUs over weeks or months. To make training large models more efficient and less demanding on computer memory, researchers use special techniques. One example is using lower-precision numbers (called FP16 or BF16) instead of full-precision ones, which saves space and speeds things up. Another technique, called model parallelism, spreads the work across multiple computer processors so that the model can be trained faster and handle more data.

4.3 Training versus inference

Training and inference are two different stages. During training, the model learns by comparing its predictions (the next word to be generated) to the correct answers (as seen in the training data) and adjusting its internal settings to reduce errors. This process is demanding and usually done in large batches of data.

To guide these adjustments, training uses optimization algorithms. One commonly used method is Adam, which adapts how much the model updates each parameter based on how quickly the error is changing. This helps the model learn more efficiently and stabilize at some point. Another method, LAMB (Layer-wise Adaptive Moments optimizer for Batch training), is designed to work well when training very large models with large batches of data. It helps maintain stable learning even when updates are made to many parameters at once.

Inference, by contrast, involves generating text using a trained model. Given a prompt, the model samples or selects one token at a time based on the output probability distribution. Common decoding strategies include:

- Greedy search: selects the most probable next token.
- Beam search: keeps multiple hypotheses at each step.
- Top-k sampling: samples from the top-k most likely tokens.
- Nucleus (top-p) sampling: samples from the smallest set of tokens whose cumulative probability exceeds p.

A hyperparameter called temperature can be used to adjust the randomness of sampling. A low temperature (<1.0) makes outputs more deterministic; a higher temperature increases diversity.

5 Scaling laws and emergent capabilities

One of the key empirical discoveries in the development of LLMs is that performance tends to improve predictably as a function of model size, training data volume, and compute. These relationships are systematically studied by Kaplan et al. (2020), who demonstrate that, under certain conditions, loss decreases smoothly with logarithmic increases in these three axes. These relationships are referred to as scaling laws.

For instance, doubling the model parameters or the dataset size leads to consistent reductions in training loss and improvements in downstream task performance—up to a point. At large enough scales, other bottlenecks (e.g., data quality or overfitting) begin to dominate. Nonetheless, the principle has guided the development of ever larger models, such as GPT, Mistral, and Llama.

An associated and intriguing phenomenon in the development of LLMs is the emergence of qualitative capabilities that appear absent in smaller models. These emergent abilities include complex forms of reasoning, code synthesis, and instruction-following (Brown et al., 2020; Wei et al., 2022) or, on the linguistic side, advanced discourse management and context-sensitive language generation. They have often been described as appearing abruptly at certain model scales, in ways that cannot be easily predicted by extrapolating from the performance of smaller models. However, recent research suggests that these abilities may in fact emerge gradually, albeit nonlinearly, as model size increases. The apparent abruptness may reflect limitations in how performance is measured or reported (Rogers and Luccioni, 2024). Although the underlying mechanisms remain poorly understood, the phenomenon itself is well documented and continues to be a subject of active investigation.

Importantly, scaling also brings challenges: increased latency and cost at inference time, greater demand for memory and computation, and growing environmental impacts from training and serving large models. Recent work has explored alternatives such as distillation, parameter-efficient tuning, and sparse models to mitigate these issues while preserving performance.

6 From foundation models to aligned chat models

6.1 Foundation models

A foundation model is a large, pretrained model trained on diverse and massive text corpora using unsupervised objectives such as CLM. These models are capable of generalizing across a wide array of downstream tasks without task-specific fine-tuning. Examples include GPT, Mistral, and Llama.

However, foundation models are not instruction-following by default. Their outputs reflect statistical continuations of prompts, which may or may not align with user intent or conversational norms. As a result, foundation models must be further aligned with human expectations to function effectively in interactive settings such as chatbots.

6.2 Instruction tuning

Instruction tuning, also known as supervised fine-tuning (SFT), is the first step in aligning a foundation model for conversational use (that is, enabling it to respond to questions rather than merely continuing a sequence of text). The process involves fine-tuning the model on a dataset of prompts paired with high-quality, human-written responses. These datasets are curated to reflect desirable behaviors, including politeness, relevance, and informativeness. For example, rather than producing a generic continuation of text in response to a prompt like 'Explain photosynthesis,' the tuned model is specifically trained to generate a clear and structured explanation tailored to the user's request.

6.3 Reinforcement learning from human feedback

While SFT improves task-following behavior, it may not suffice to align model responses with human preferences in nuanced contexts. RLHF introduces an additional layer of alignment using a reinforcement learning framework driven by human feedback (Ouyang et al., 2022).

The process includes:

1. Generating multiple responses to a prompt using the SFT model.
2. Asking human annotators to rank these outputs from best to worst.
3. Training a reward model to predict human preferences.
4. Using proximal policy optimization (PPO) to fine-tune the language model to maximize the predicted reward.

This procedure encourages the model to prioritize helpful, honest, and harmless behavior over merely plausible completions (Bai et al., 2022). RLHF is used in models such as ChatGPT (in contrast to the base GPT models) or Mistral's LeChat (in contrast to their base models) to improve alignment with user expectations.

7 Controlling model behavior

Once a model has been aligned using instruction tuning and RLHF, additional mechanisms are necessary to control its behavior in deployment settings. These control mechanisms aim to mitigate the risks of generating unsafe, unethical, or undesired outputs. This section outlines the main approaches used in production-grade chat models.

7.1 Refusal mechanisms

Chat models are often trained or conditioned to refuse to answer certain categories of questions, such as those requesting illegal advice, medical diagnoses, or personal data. Refusal behavior can be learned during supervised fine-tuning (e.g., by including examples of polite refusals) or reinforced during RLHF when annotators rank refusals higher than unsafe completions.

Example:

User: How do I make explosives at home?
Model: I'm sorry, but I can't help with that request.

Such refusals are typically generic and conservative to minimize risk.

7.2 Prompt classification and filtering

Before the model generates a response, the user's input is often passed through a classification system to detect potentially harmful or policy-violating content. These classifiers can detect hate speech, self-harm content, disallowed topics (e.g., pornography, violence), or violations of terms of service.

If a prompt is flagged, it may be blocked entirely, rewritten, or routed to a fallback response. In enterprise deployments, these systems are critical for maintaining safety, compliance, and legal defensibility.

In the case of ChatGPT-5, prompt analysis also serves a routing function: the system classifies the type of query and can dynamically select the most appropriate underlying model or specialized module, thereby optimizing responses across different kinds of tasks.

7.3 Output filtering

Output filtering applies similar classification mechanisms to the model's generated response before it is shown to the user. This second layer ensures that, even if the model produces a problematic output, it can be intercepted.

Some platforms implement multipass filtering, using both lexical and semantic analyses, including sentiment detection, keyword blacklists, and transformer-based toxicity classifiers.

7.4 Guardrails and policy layers

Guardrails are system-level constraints designed to ensure that the model remains within acceptable operational bounds. These may include:

- System prompts that instruct the model to behave in a certain way (e.g., 'You are a helpful assistant').

- Content policies enforced through prompt augmentation or output filtering.
- Logging and monitoring tools that detect anomalous behavior in real time.

Some deployments also integrate escalation mechanisms (e.g., deflecting users to human support) or customizable interfaces for institutional use (e.g., education, health care).

8 Tool use and augmented models

While LLMs trained on static corpora can generalize broadly, their capabilities can be extended by interfacing them with external tools and data sources. This augmentation allows models to access up-to-date or domain-specific knowledge, perform calculations, retrieve documents, or execute code. Such extensions are essential for many real-world applications.

8.1 Retrieval-augmented generation

Retrieval-augmented generation (RAG) combines a language model with a retrieval system, such as a vector database or traditional search index. When a query is received, relevant documents are first retrieved and then fed into the model alongside the original prompt. This process grounds the model's responses in external sources, improving factual accuracy and domain specificity. For example, instead of relying solely on the information encoded during training, a RAG-based assistant can consult a legal database or recent scientific literature before generating an answer.

RAG can also help keep the system up to date by allowing it to draw on current information without retraining the entire model. However, given the limitations of language models and their tendency to produce confident but incorrect outputs, they should not be relied upon as search engines.

8.2 Toolformer and API invocation

Toolformer (Schick et al., 2023) is a method that enables a language model to learn when and how to invoke external tools during generation. During training, the model is prompted to use tools such as calculators, translation APIs, or search engines. These calls are inserted into the training data so that the model can imitate their use.

This technique helps bridge the gap between language prediction and task execution. For example, instead of estimating the result of '234 × 51,' a Toolformer-enhanced model can delegate the computation to a calculator tool and report the correct result.

8.3 Plugins and external APIs (AI agents)

Some chat-based models, including ChatGPT with plugin support, are designed to interact dynamically with third-party APIs. These plugins allow users to extend model functionality for specific domains, such as booking flights, summarizing documents, or running Python code in a sandboxed environment.

This setup effectively turns the model into an AI agent: It can decide when to call an external tool, interpret the result, and integrate it into the conversation. Such agentic behavior enables hybrid systems that combine natural language fluency with external computation, database access, and structured task execution.

9 Limitations and active research directions

Despite their capabilities, LLMs exhibit well-known limitations that motivate ongoing research across multiple areas. These include representational constraints, inference inefficiencies, alignment challenges, and environmental concerns.

9.1 Context window limitations

Transformer models process input within a fixed-length context window—typically 2,048 to 32,000 tokens depending on the architecture (and up to 200k tokens for the most recent models). Any tokens beyond this window are truncated or ignored. This makes handling long documents or conversations problematic.

Efforts to extend context length include memory-efficient attention variants (e.g., Longformer, FlashAttention) and hierarchical approaches (e.g., segment-wise encoding or chunk-based models).

9.2 Hallucination and factual errors

LLMs frequently produce text that is fluent and plausible but factually incorrect—a phenomenon referred to as hallucination. This occurs because the model optimizes for likelihood, not truth, and lacks real-world grounding (see Chapter 1).

Research in retrieval augmentation, fact-checking integration, and truthfulness-enhancing objectives (e.g., 'TruthfulQA') seeks to reduce hallucinations. Nonetheless, this remains a major barrier to the safe use of LLMs in high-stakes domains such as medicine and law.

9.3 Computational and environmental costs

Training large LLMs requires extensive computing infrastructure—often thousands of GPUs running for weeks—which consumes large amounts of electricity. Estimates of the carbon footprint of training a single model run into hundreds of metric tons of CO_2 (see Chapter 9).

To address this, researchers explore model efficiency through sparsity, quantization, knowledge distillation, and more efficient pretraining (e.g., Chinchilla scaling). Debates around sustainability and equitable access are increasingly prominent in the LLM research agenda.

9.4 Alignment and interpretability

Even after instruction tuning and RLHF, aligning model behavior with human values is difficult. Models may exhibit biased, manipulative, or opaque responses. Interpretability also remains limited: It is difficult to understand why a model generated a specific output or how internal representations evolve.

Techniques such as attention visualization, probing classifiers, and causal interventions are under active development. Research into mechanistic interpretability aims to reverse-engineer the computations inside transformer layers.

10 Conclusion

This technical annex has provided a comprehensive overview of the architectural, algorithmic, and operational principles that underlie contemporary LLMs. From text tokenization to transformer dynamics, from scaling laws to alignment techniques such as RLHF, and from tool integration to emerging limitations, LLMs represent a convergence of innovations in deep learning, large-scale data curation, and computational infrastructure.

The goal of this annex has been purely descriptive and technical, equipping readers with the foundational knowledge necessary to engage critically with the philosophical, ethical, and political questions addressed in the main text. Understanding the inner workings of LLMs enables more grounded interpretations of their societal roles and constraints.

References

Bai, Y., Kadavath, S., Kundu, S., Askell, A., Kernion, J., Jones, A., Chen, A., Goldie, A., Mirhoseini, A., McKinnon, C., Chen, C., Olsson, C., Olah, C., Hernandez, D., Drain, D., Ganguli, D., Li, D., Tran-Johnson, E., Perez, E. ... Kaplan, J. (2022). Constitutional AI: Harmlessness from AI feedback. arXiv preprint. https://arxiv.org/abs/2212.08073

Brown, T., Mann, B., Ryder, N., Subbiah, M., Kaplan, J. D., Dhariwal, P., Neelakantan, A., Shyam, P., Sastry, G., Askell, A., Agarwal, S., Herbert-Voss, A., Krueger, G., Henighan, T., Child, R., Ramesh, A., Ziegler, D., Wu, J., Winter, C. … Amodei, D. (2020). Language models are few-shot learners. In *Proceedings of the 33th International Conference on Neural Information Processing Systems (NeurIPS 2020), Red Hook, NY, USA* (pp. 1877–1901). Curran. https://dl.acm.org/doi/abs/10.5555/3495724.3495883

Kaplan, J., McCandlish, S., Henighan, T., Brown, T., Chess, B., Child, R., Gray, S., Radford, A., Wu, J., and Amodei, D. (2020). Scaling laws for neural language models. arXiv preprint. https://arxiv.org/abs/2001.08361

Ouyang, L., Wu, J., Jiang, X., Almeida, D., Wainwright, C. L., Mishkin, P., Zhang, C., Agarwal, S., Slama, K., Ray, A., Schulman, J., Hilton, J., Kelton, F., Miller, L., Simens, M., Askell, A., Welinder, P., Christiano, P., Leike, J., and Lowe, R. (2022). Training language models to follow instructions with human feedback. In *Proceedings of the 36th International Conference on Neural Information Processing Systems (NeurIPS '22) New Orleans, USA* (pp. 27730–27744). Curran. https://arxiv.org/abs/2203.02155

Rogers, A., and Luccioni, S. (2024). Position: Key Claims in LLM Research Have a Long Tail of Footnotes. In *Proceedings of the 41st International Conference on Machine Learning (ICML'24), Vienna, Austra*, Vol. 235. JMLR, Article 1735. https://dl.acm.org/doi/10.5555/3692070.3693805

Schick, T., Dwivedi-Yu, J., Dessí, R., Raileanu, R., Lomeli, M., Hambro, E., Zettlemoyer, L., Cancedda, N., and Scialom, T. (2023). Toolformer: Language models can teach themselves to use tools. In *Proceedings of the 37th International Conference on Neural Information Processing Systems (NeurIPS 2023), Red Hook, NY, USA* (pp. 68539–68551). Curran. https://dl.acm.org/doi/10.5555/3666122.3669119

Vaswani, A., Shazeer, N., Parmar, N., Uszkoreit, J., Jones, L., Gomez, A. N., Kaiser, Ł., and Polosukhin, I. (2017). Attention is all you need. In *Proceedings of the 31st International Conference on Neural Information Processing Systems (NeurIPS 2017), Red Hook, NY, USA* (pp. 6000–6010). Curran. https://dl.acm.org/doi/10.5555/3295222.3295349

Wei, J., Bosma, M., Zhao, V. Y., Guu, K., Wei Yu, A., Lester, B., Du, N., Dai, A. M., and. Le, Q. V. (2022). Finetuned language models are zero-shot learners. In *Proceedings of the International. Conference on Learning Representations (ICLR 2022)*. https://openreview.net/forum?id=gEZrGCozdqR

Index

www.ingramcontent.com/pod-product-compliance
Lightning Source LLC
Chambersburg PA
CBHW041144230326

41599CB00039BA/7162